ELECTROMAGNETICS FOR PRACTICING ENGINEERS

SECOND EDITION

For a complete listing of titles in the
Artech House Electromagnetic Analysis Library,
turn to the back of this book.

ELECTROMAGNETICS FOR PRACTICING ENGINEERS

SECOND EDITION

Dean James Friesen

ARTECH HOUSE
BOSTON | LONDON
artechhouse.com

Library of Congress Cataloging-in-Publication Data
A catalog record for this book is available from the U.S. Library of Congress.

British Library Cataloguing in Publication Data
A catalogue record for this book is available from the British Library.

ISBN 13: 978-1-68569-149-3

Cover design by Joi Garron

Artech House
685 Canton Street
Norwood, MA 02062

10 9 8 7 6 5 4 3 2 1

To my faithful, loyal brother Gregg

CONTENTS

6

CURRENT, CURRENT DENSITY, AND CONDUCTORS 73

7

CAPACITANCE AND DIELECTRIC MATERIALS 93

8

LAPLACE'S EQUATION 113

Part 2

MAGNETOSTATICS: BASIC PRINCIPLES 129

9

AMPERE'S LAW AND THE MAGNETIC FIELD 131

10

FORCES AND TORQUES IN MAGNETIC FIELDS 153

11

INDUCTANCE AND MAGNETIC CIRCUITS 171

12

DISPLACEMENT CURRENT AND INDUCED EMF 191

Part 3
PRINCIPLES OF FREE-SPACE ELECTROMAGNETICS 207

13
ELECTROMAGNETIC WAVES 209

14
BOUNDARY CONDITIONS 235

15
MAXWELL'S EQUATIONS 241

16

Practical Solutions To Real-World Electromagnetic Engineering Problems 251

A

SCIENTIFIC PREFIXES 293

B

SCIENTIFIC CONSTANTS 295

PREFACE

This second edition was targeted specifically for the young professional entering the engineering industry after college or university study. Truth be told, I did not leave Kansas State University (KSU in Manhattan, KS) in 1987 with my BS degree in engineering fully adept at the subject of electromagnetics (EM). This took some initiative on my part, years after graduation.

First, in June 1987 I was introduced to some of the finest colleagues a young professional could ever be associated with. These guys introduced me to and taught me about the field of electronic warfare, and I had three good starting years. However, in October 1991, the engineering market changed, and I found myself feeling the need to expand my engineering skills in the areas of radar (1987–1990), voice and digital radio communications (1992–2000), and electromagnetic theory (2004–2011). Good books on radar and voice/digital radio communications were easily procured. However, the topic of electromagnetics was considerably more difficult to tackle by way of self-study.

I chose not to enroll in a university or community college course. Neither did I attend a professional engineering seminar. Instead, I had the good fortune to find procure three important learning resources:

1. A clearly written university-level vector calculus book;
2. An excellently taught YouTube video course titled "Electromagnetics" (24 1-hr lectures) taught by Dr. Harishankar Ramashandran of the Indian Institute of Technology;
3. A clearly understandable Schaum's Outline series titled, *Theory and Problems of Electromagnetics* written by Joseph A. Edminister, to tackle EM problem-solutions.

It took me four years to go through the full calculus book (800 pages), an additional four months to view and completely learn the contents of the 24 1-hr Ramashandran EM lectures, and another 2 years and 2 months to tackle (and conquer) the problems offered in the Schaum's Outline book. I devoted anywhere between 15 to 45 minutes a day, even as I worked full-time on radio frequency (RF)/ microwave (MW), electromagnetic compatibility (EMC), and antenna engineering for Boeing. I eventually became so proficient in my skill and understanding of EM that colleagues with master's degrees and PhD's in the company would come to me to consult on various EM problems they encountered. In 3 months, the team's electromagnetic effects (EME) and antennas group manager promoted me to senior principal investigator for the BCA Product Development organization of Boeing (Seattle, WA).

I share this with you—not to be boastful—but to illustrate that with enough effort, patience, and perseverance, you, too, can gain proficiency and self-confidence in this subject such that real-world EM problems/obstacles can be overcome in the practical, real-world engineering arena, too.

I wish the reader every good success in his/her efforts and future professional endeavors.

PART 1

ELECTROSTATICS: BASIC PRINCIPLES

1

VECTOR ANALYSIS

Electromagnetics principles can be described adequately with precise and well-crafted words. However, in the practical engineering world, there often arises the need to quantify the magnitude of an important and unsolved electromagnetic engineering problem—and effectively convey it to senior engineering and management. For such often unwelcome occasions, vector calculus comes in handy, seeing how it helped form (by way of many predecessors such as James Clerk Maxwell), the foundation of electromagnetics, more specifically, electrostatics and magnetostatics. Vector calculus is indispensable to the self-study and successful acquisition of a thorough understanding of the subject(s).

For convenience, a short refresher on vector calculus is offered in this chapter. The author nonetheless advises readers to acquire for themselves a very good and easy-to-comprehend calculus book and emphasizes using it to strengthen the reader's skill. It will make understanding the underlying principles of electromagnetics far easier. This simple overview also aids in keeping the size of this first volume to a manageable level for the publisher. All said, a highly recommended university/college-level, self-study calculus textbook is cited in [1].

1.1 INTRODUCTION

Discussion in this first chapter begins with a brief tutorial on vectors. Relevant coordinate systems are presented as are discussions on vector notation and structure, and the parallelogram and head-to-tail rules (see Appendix C). The unit vector is presented as: (1) differential volume, (2) differential surface (area), and (3) differential line elements. Discussion on vector fields and coordinate system transformations are also given.

1.2 VECTOR NOTATION

It is vital for the engineering analyst to be able to distinguish between vector quantities and scalar quantities. Vectors are quantities that have both a magnitude and a direction. Scalars are quantities that have only a magnitude; that is, no indicated direction [1]. Vectors throughout this text are indicated by boldface type. Scalars are denoted in plain type.

Distance formula is the distance between two points in three-dimensional space (Cartesians coordinates). It is a scalar value that can be found, given two coordinate points in space, by the following formula:

$$D = \left[(x_1 - x_0)^2 + (y_1 - y_0)^2 + (z_1 - z_0)^2\right]^{1/2} \tag{1.1}$$

where

P_0 (x_0, y_0, z_0) and P_1 (x_1, y_1, z_1) are points in space [2].

A unit vector is always a vector having an absolute value (i.e., a magnitude or length) of 1 and a direction [3]. It will always be indicated in this text by boldface, lower case letters, such as, **a** or **u**. The unit vector in the direction of a vector **A** is determined by dividing **A** by its absolute value (or vector length) as shown below:

$$\mathbf{a}_A = \frac{\mathbf{A}}{|\mathbf{A}|} \text{ or } \frac{\mathbf{A}}{A} \tag{1.2}$$

where

$|\mathbf{A}| = A = (A_x^2 + A_y^2 + A_z^2)^{1/2}$

A_x is the magnitude of the **A** vector in the x-direction

A_y is the magnitude of the **A** vector in the y-direction

A_z is the magnitude of the **A** vector in the z-direction

It should be stated that many calculus texts use the boldfaced variable **u** to denote the unit vector as well [2].

Using unit vectors, $\mathbf{a}_x$, $\mathbf{a}_y$, and $\mathbf{a}_z$ along the Cartesian x, y, and z system coordinate axes, an arbitrary vector can be written in component form as:

$$\mathbf{A} = A_x\mathbf{a}_x + A_y\mathbf{a}_y + A_z\mathbf{a}_z = A_x\mathbf{x} + A_y\mathbf{y} + A_z\mathbf{z} \tag{1.3}$$

However, for writing convenience and reduction of confusion, the above-expressed Cartesian-based unit vectors will be written in this text from here forward as **x**, **y**, and **z**, as opposed to $\mathbf{a}_x$, $\mathbf{a}_y$, and $\mathbf{a}_z$. The same parallel cases will also be for unit vectors for cylindrical and spherical coordinates [3].

Hence, unit vectors for the cylindrical coordinate system will be written as the more conveniently written

$$\mathbf{r},\ \boldsymbol{\phi},\ \text{and}\ \mathbf{z}$$

instead of the more awkward mathematician's form of $\mathbf{a}_r$, $\mathbf{a}_\phi$, and $\mathbf{a}_z$, respectively. Likewise, the case for the spherical coordinate system; that is:

$$\mathbf{r},\ \boldsymbol{\phi},\ \text{and}\ \boldsymbol{\theta}$$

for $\mathbf{a}_r$, $\mathbf{a}_\phi$, and $\mathbf{a}_\theta$ [1]. In determining the resultant vector $\mathbf{A} + \mathbf{B}$, that is in the case where two vectors follow each other in a head-to-tail connection with each other, the parallelogram rule is frequently employed. Because it is important to understand well the parallelogram rule for the solution of many electrostatics problems, a detailed discussion is offered in Appendix C of this book.

1.3 VECTOR ALGEBRA

1. Vectors may be added and subtracted.

$$\begin{aligned}\mathbf{A} \pm \mathbf{B} &= \left(A_x\mathbf{x} + A_y\mathbf{y} + A_z\mathbf{z}\right) \pm \left(B_x\mathbf{x} + B_y\mathbf{y} + B_z\mathbf{z}\right) \\ &= \left(A_x \pm B_x\right)\mathbf{x} + \left(A_y \pm B_y\right)\mathbf{y} + \left(A_z \pm B_z\right)\mathbf{z}\end{aligned} \tag{1.4}$$

2. The associative,

$$\mathbf{A} + (\mathbf{B} + \mathbf{C}) = (\mathbf{A} + \mathbf{B}) + \mathbf{C} \tag{1.5}$$

distributive,

$$k(\mathbf{A} + \mathbf{B}) = k\mathbf{A} + k\mathbf{B}, \quad (k_1 + k_2)\mathbf{A} = k_1\mathbf{A} + k_2\mathbf{A} \tag{1.6}$$

where

k is an arbitrary constant of any value

k_1 is an arbitrary constant of a particular value

k_2 is a second arbitrary constant of a value not equal to k_1

and commutative laws apply.

$$\mathbf{A} + \mathbf{B} = \mathbf{B} + \mathbf{A} \tag{1.7}$$

3. The dot product of two vectors is, by definition:

$$\mathbf{A} \cdot \mathbf{B} = |\mathbf{A}|\,|\mathbf{B}|\cos\Theta = AB\cos\Theta \quad \text{(Note: This reads “A dot B”)} \tag{1.8}$$

where Θ is the smaller angle between vectors **A** and **B**.

From the component form it can be shown that:

$$\mathbf{A} \cdot \mathbf{B} = A_x B_x + A_y B_y + A_z B_z \tag{1.9}$$

in particular,

$$\mathbf{A} \cdot \mathbf{A} = |\mathbf{A}|^2 = A_x^2 + A_y^2 + A_z^2 \tag{1.10}$$

4. The cross product of two vectors is, by definition:

$$\mathbf{A} \times \mathbf{B} = (|\mathbf{A}||\mathbf{B}|\sin\theta)\mathbf{n} = (AB\sin\Theta)\mathbf{n} \quad \text{(Reads “A cross B”)} \tag{1.11}$$

where Θ is the smaller angle between vectors **A** and **B** and **n** is a unit vector normal to the plane determined by **A** and **B** when they are drawn from a point common to both [1].

In the cross product, two normals are in the plane [1], and, thus, further specification is required [4]. The selected normal,

$$\mathbf{A} \times \mathbf{B} = -\mathbf{B} \times \mathbf{A} \tag{1.12}$$

is the one that is in the direction of a right-hand screw when **A** is turned toward **B** (see Figure 1.1).

The cross product is anticommutative; that is, it does not follow the commutative principle [1]. Expanding the cross product into component form yields:

$$\begin{aligned}\mathbf{A} \times \mathbf{B} &= \left(A_x\mathbf{x} + A_y\mathbf{y} + A_z\mathbf{z}\right) \times \left(B_x\mathbf{x} + B_y\mathbf{y} + B_z\mathbf{z}\right)\\ &= \left(A_yB_z - A_zB_y\right)\mathbf{x} + \left(A_zB_x - A_xB_z\right)\mathbf{y} + \left(A_xB_y - A_yB_x\right)\mathbf{z}\\ &= \begin{bmatrix} \mathbf{x} & \mathbf{y} & \mathbf{z} \\ A_X & A_y & A_z \\ B_x & B_Y & B_z \end{bmatrix}\end{aligned} \tag{1.13}$$

conveniently expressed in determinant form (< immediately left).

The following is a triple scalar product of three vectors (**A**, **B**, and **C**):

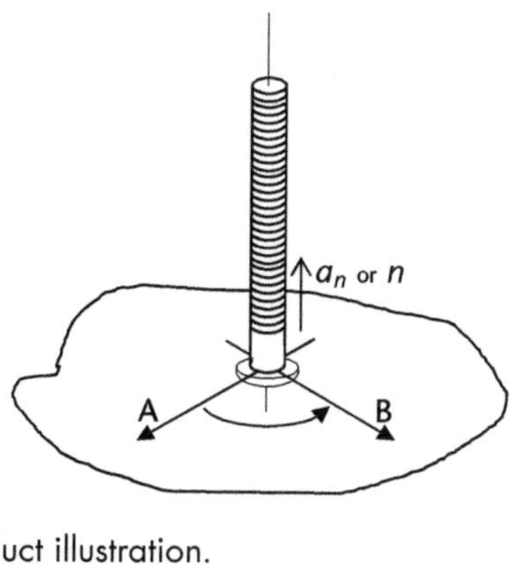

Figure 1.1 Cross-product illustration.

$$[\mathbf{A}\ \mathbf{B}\ \mathbf{C}] = \mathbf{A} \cdot (\mathbf{B} \times \mathbf{C}) = (\mathbf{A} \times \mathbf{B}) \cdot \mathbf{C} = \begin{bmatrix} A_1 & A_2 & A_3 \\ B_1 & B_2 & B_3 \\ C_1 & C_2 & C_3 \end{bmatrix} \tag{1.14}$$

Equation (1.14) yields the volume of a parallelepiped (see Figure 1.2) [1].

1.4 COORDINATE SYSTEMS

The Cartesian coordinate system, displayed in Figure 1.3(a), is one of three coordinate systems that are used to help solve electromagnetics-related engineering problems. The other two coordinate systems are the cylindrical coordinate system (see Figure 1.3(b)) and the spherical coordinate system (see Figure 1.3(c)) [1]. All three are displayed in the figure to give the reader the chance to visualize and understand similarities/differences between the coordinate systems.

As can be seen in Figure 1.3, the Cartesian and cylindrical coordinate systems have the z-axis in common. The cylindrical and spherical coordinate systems have the ϕ-coordinate in common. However, while these two systems also have an r-coordinate, each

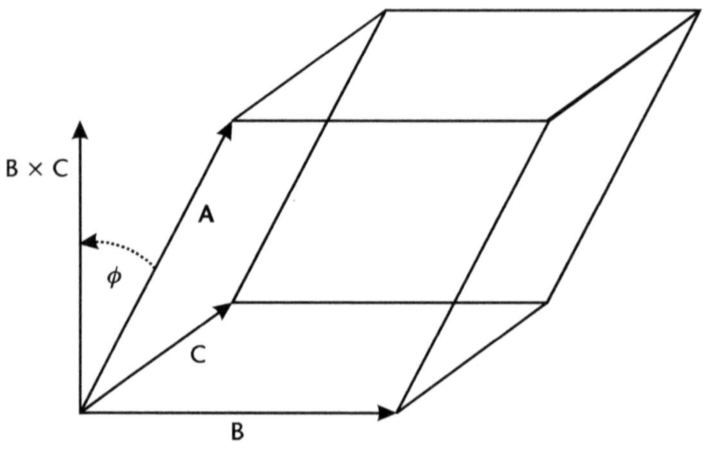

$$V_{parallelepiped} = |\mathrm{A} + (\mathrm{B} \times \mathrm{C})|$$
$$= |\mathrm{B} \times \mathrm{C}|\ |\mathrm{A}|\ |\cos \phi$$

Figure 1.2 Parallelepiped illustration with associated vector expressions.

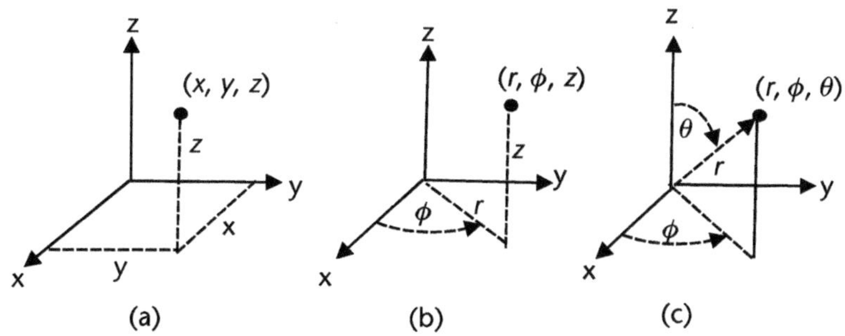

Figure 1.3 Three coordinate systems that are used to solve electromagnetic problems: (a) Cartesian, (b) cylindrical, and (c) spherical.

manifests itself in slightly different orientations. Hence, the *r*-coordinate for the spherical coordinate system appears to have slightly more directional freedom than the *r*-coordinate for the cylindrical coordinate.

Figure 1.4 provides a second helpful set of illustrations for all three coordinate systems:

- Figure 1.4(a)—the Cartesian coordinate system;
- Figure 1.4(b)—the cylindrical coordinate system;
- Figure 1.4(c)—the spherical coordinate system.

Each of the illustrations intends to show the geometric envelope of each system [1].

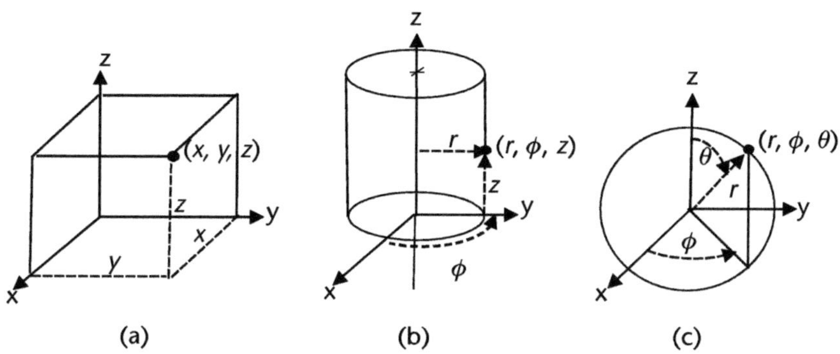

Figure 1.4 Illustrations of the use for each three-dimensional coordinate system: (a) Cartesian, (b) cylindrical, and (c) spherical.

Ranges of the coordinates (r, ϕ, z) in the cylindrical coordinate system are:

$$0 < r < \infty, \qquad 0 < \phi < 2\pi, \qquad -\infty < z < +\infty \tag{1.15}$$

Ranges of the coordinates (r, ϕ, θ) in the spherical coordinate system are:

$$0 < r < \infty, \qquad 0 < \phi < 2\pi, \qquad 0 < \theta < \pi \tag{1.16}$$

The Cartesian coordinate system, the one shown in Figures 1.3(a) and 1.4(a), is the one with which beginning engineers and engineering students are likely to have the greatest initial familiarity. However, in certain cases, solutions are sometimes found to be easier to obtain by using either the cylindrical or spherical coordinate system rather than the Cartesian [1].

Practical, real-world examples for the use of the cylindrical coordinate system include the geometries of wires, wire bundle harnesses, coaxial cables, and airplane fuel quantity sensors. In these cases, the cylindrical coordinate system lends itself more easily to solution-finding than does the Cartesian coordinate system. For the study of antennas in antenna reference texts, the reader will find that antenna performance and antenna patterns (E-field or power) are more easily evaluated in the spherical coordinate system.

In this text, the right-handed versions of the Cartesian, the cylindrical, and the spherical coordinate systems are used. (Why add to the confusion to use the left-handed version? It adds no value to the real-world of finding solutions to practical problems.) Hence:

$$\mathbf{x} \times \mathbf{y} = \mathbf{z} \tag{1.17}$$

$$\mathbf{r} \times \boldsymbol{\phi} = \mathbf{z} \tag{1.18}$$

$$\mathbf{r} \times \boldsymbol{\theta} = \boldsymbol{\phi} \tag{1.19}$$

The component forms of a vector, $\mathbf{A}$, in the three coordinate systems are:

$$\mathbf{A} = A_x\mathbf{x} + A_y\mathbf{y} + A_z\mathbf{z} \quad \text{(Cartesian)} \tag{1.20}$$

$$\mathbf{A} = A_r\mathbf{r} + A_\phi\boldsymbol{\phi} + A_z\mathbf{z} \quad \text{(cylindrical)} \tag{1.21}$$

$$\mathbf{A} = A_r\mathbf{r} + A_\phi\boldsymbol{\phi} + A_\theta\boldsymbol{\theta} \quad \text{(spherical)} \tag{1.22}$$

Most often, the components A_x, A_y, A_z, A_r, A_ϕ, and A_θ shown in (1.20) to (1.22), generally speaking, are not always constants. It is important to understand that each can also be *functions* in the coordinate system form as well [1].

1.5 DIFFERENTIAL VOLUME, SURFACE, AND LINE ELEMENTS

Differential Volume. When the coordinates of a point P are expanded to:

$$(x + dx, y + dy, z + dz) \quad \text{(Cartesian coordinate system)} \tag{1.23}$$

$$(r + dr, \phi + d\phi, z + dz) \quad \text{(cylindrical coordinate system)} \tag{1.24}$$

or

$$(r + dr, \theta + d\theta, \phi + d\phi) \quad \text{(spherical coordinate system)} \tag{1.25}$$

a differential volume, dV, is formed [1]. The expressions for each coordinate system are:

$$dV = dx\ dy\ dz \quad \text{(Cartesian coordinate system)} \tag{1.26}$$

$$dV = dr(r\ d\phi)dz = r\ dr\ d\phi\ dz \quad \text{(cylindrical coordinate system)} \tag{1.27}$$

$$dV = dr(r\ d\theta)(r\sin\theta\ d\phi) = r^2\ dr\sin\theta\ d\theta\ d\phi \quad \text{(spherical coordinate system)} \tag{1.28}$$

To the first order in infinitesimal quantities, the differential volume for the Cartesian coordinate system is a rectangular box (i.e., cuboid). See Figure 1.5(a) [1]. In the other two coordinate systems (see Figures 1.5(b) and 1.5(c)), the corresponding volumes are slightly distorted variations from the Cartesian cuboid.

The reader is encouraged to carefully study the latter two differential volumes until a good, accepted understanding of each is reached.

Differential surface (area). When a differential surface (area) is the focus for quantitative evaluation, the dS expressions for each of the three coordinate systems are used [6]. These are presented for convenience as follows:

$$dS = dx\ dy \quad dS = dy\ dz \quad dS = dx\ dz \quad \text{(Cartesian)} \tag{1.29}$$

$$\begin{aligned} dS = dr\ dz \quad dS &= dz(r\ d\phi) \quad dS = dr(r\ d\phi) \\ &= r\ d\phi\ dz \qquad\quad = (r\ dr)d\phi \end{aligned} \quad \text{(cylindrical)} \tag{1.30}$$

$$\begin{aligned} dS &= dr(r\ d\theta) \quad dS = dr(r\sin\theta\ d\phi) \quad dS = (r\ d\theta)(r\sin\theta\ d\phi) \\ &= r\ dr\ d\phi \qquad\qquad\qquad\qquad\qquad\quad = r^2\sin\theta\ d\theta\ d\phi \end{aligned} \quad \text{(spherical)} \tag{1.31}$$

The reader is urged to also use Figure 1.5 as an aid to gaining visual understanding of differential surfaces on each of the differential volumes as well.

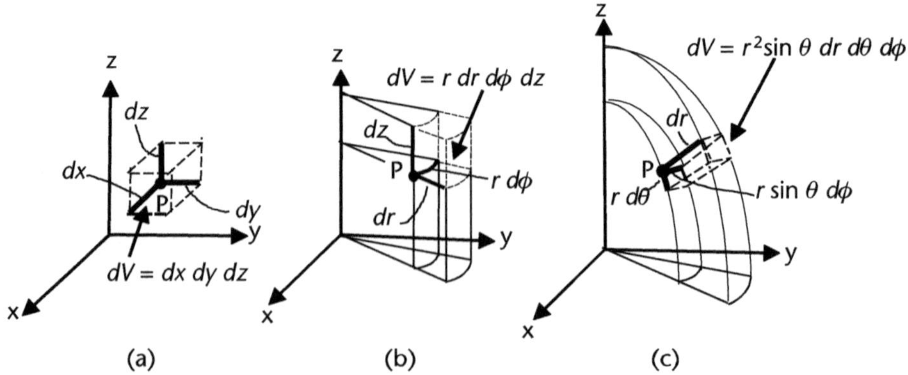

Figure 1.5 Differential volume, dV, in each of three coordinate systems: (a) Cartesian, (b) cylindrical, and (c) spherical.

Differential line element. The differential line, dl, is the diagonal through an arbitrary point P [5].

Expressions for the differential line elements that correspond to each of the coordinate systems are given as follows [5]:

$$dl = dx^2 + dy^2 + dz^2 \qquad \text{(Cartesian)} \tag{1.32}$$

$$dl = dr^2 + r^2 d\phi^2 + dz^2 \qquad \text{(cylindrical)} \tag{1.33}$$

$$dl = dr^2 + r^2 d\theta^2 + r^2 \sin^2\theta d\phi^2 \qquad \text{(spherical)} \tag{1.34}$$

It is vital for the reader to understand that the above formulas for differential volume, differential surface, and differential line element will be fundamental to quantitative problem-solving in Chapters 2 to 8.

1.6 VECTOR FIELDS

A *field* is a spatial distribution of a quantity that may or may not be a function of time. Generally speaking, coefficients of electromagnetic vector expressions contain variables. As such, the expression changes magnitude and direction from point to point throughout the region of interest [1]. For example, consider the vector:

$$\mathbf{E} = -x\mathbf{x} + y\mathbf{y} \tag{1.35}$$

Values of x and y may be substituted into the expression to give $\mathbf{E}$ at the various locations.

After a number of points are examined, the pattern becomes evident, and Figure 1.6 shows this field.

A vector field may also vary with time, and, thus, the two-dimensional field examined in Figure 1.6 may also change with time. It could be assigned a time-variation such as:

$$\mathbf{E} = \left(-x\mathbf{a}_\mathrm{x} + y\mathbf{a}_\mathrm{y}\right)\sin(\omega t) \ \text{ or } \ E = \left(-x\mathbf{a}_\mathrm{x} + y\mathbf{a}_\mathrm{x}\right)e^{j\omega t} \tag{1.36}$$

The electric and magnetic fields of Chapter 2 and beyond are all time-variable. As might be expected, they will need to be

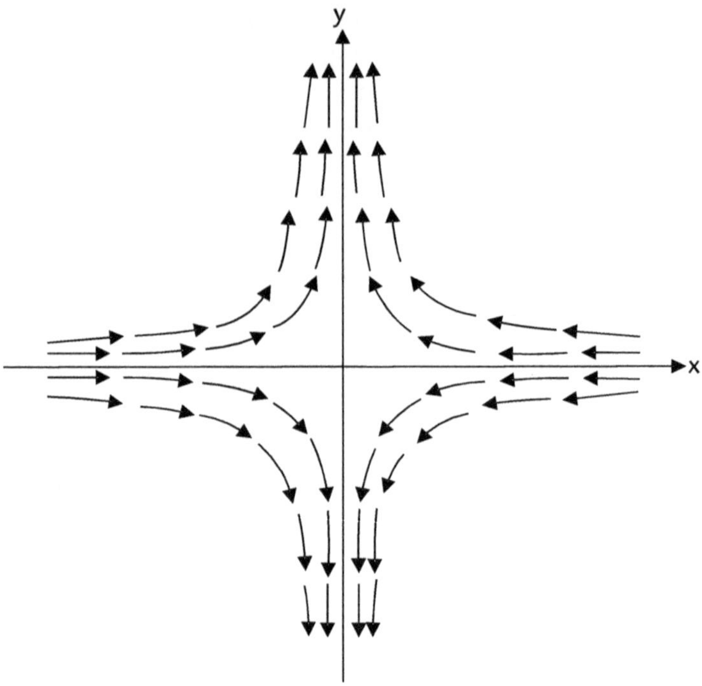

Figure 1.6 Example of a vector field, illustrated.

differentiated with respect to time and integrated with respect to time as the solution situation dictates. However, in time the reader will find that both operations follow naturally and rarely cause great difficulty [1].

If the reader believes he or she would benefit from vector analysis practice, he or she is encouraged to go to Section 1.8 where sample vector analysis problems are performed.

1.7 TRANSFORMATIONS BETWEEN COORDINATE SYSTEMS

Sometimes, it is helpful to transform points in space to one or the other of the two coordinate systems in order to obtain a solution more quickly. Here are transformations [1]:

Cartesian to cylindrical coordinate system. To deal with problems having cylindrical electromagnetic structures, the cylindrical coordinate system is used. If the reader finds it helpful to transform Cartesian coordinate points in space to equivalent cylindrical coordinate points, then the following transformational equations are used:

$$r = \left(x^2 + y^2\right)^{1/2} \qquad \phi = \arctan\left(\frac{y}{x}\right) \qquad z = z \tag{1.37}$$

Reverse cylindrical to Cartesian transformational equations are:

$$x = r\cos\phi \qquad y = r\sin\phi \qquad z = z \tag{1.38}$$

Cartesian to spherical coordinate system. To deal with problems having spherical electromagnetic structures, the spherical coordinate system is used. If the reader finds it helpful to transform a Cartesian coordinate point to an equivalent spherical coordinate point, then the following transformational equations are employed:

$$r = \left(x^2 + y^2 + z^2\right)^{1/2} \qquad \theta = \arctan\left[\frac{z}{\left(x^2 + y^2 + z^2\right)^{1/2}}\right] \qquad \phi = \arctan\left(\frac{y}{x}\right) \tag{1.39}$$

Reverse spherical to Cartesian transformational equations are:

$$x = r\sin\theta\cos\phi \qquad y = r\sin\theta\sin\phi \qquad z = r\cos\theta \tag{1.40}$$

1.8 PROBLEMS AND SOLUTIONS: VECTOR ANALYSIS

Problem 1.1

See Figure 1.7. Find the expression for vector **C** that is directed from point Q_1 (x_1, y_1, z_1) to Q_2 (x_2, y_2, z_2).

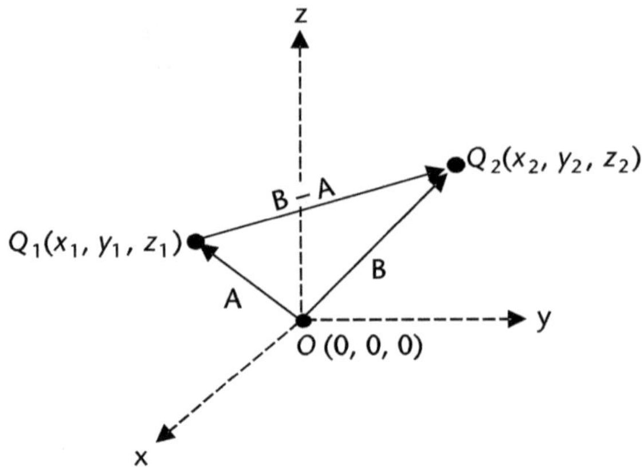

Figure 1.7 Vector diagram illustration for Problem 1.1.

Problem 1.2

See Figure 1.8. Find:

(a) The expression for vector **A** when Q_1 is located at (2, −4, 1) to Q_2 is located at (0, −2, 0) in Cartesian coordinates;

(b) The unit vector along vector **A**.

Problem 1.3

See Figure 1.9. Find the distance between the points P_0[0, −5, 0] and P_1(0, 5, 10) in Cartesian coordinates.

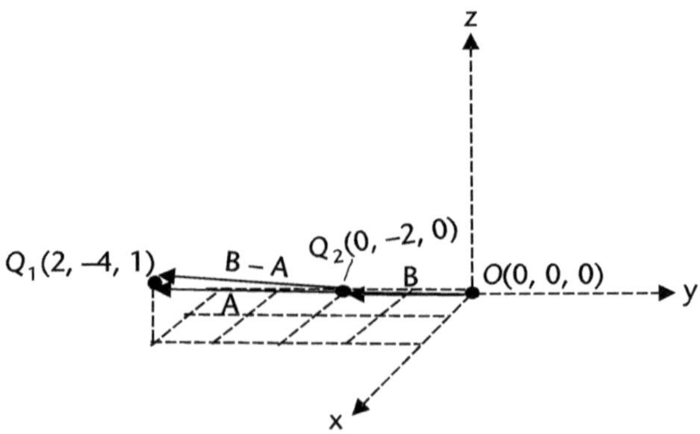

Figure 1.8 Vector diagram illustration for Problem 1.2.

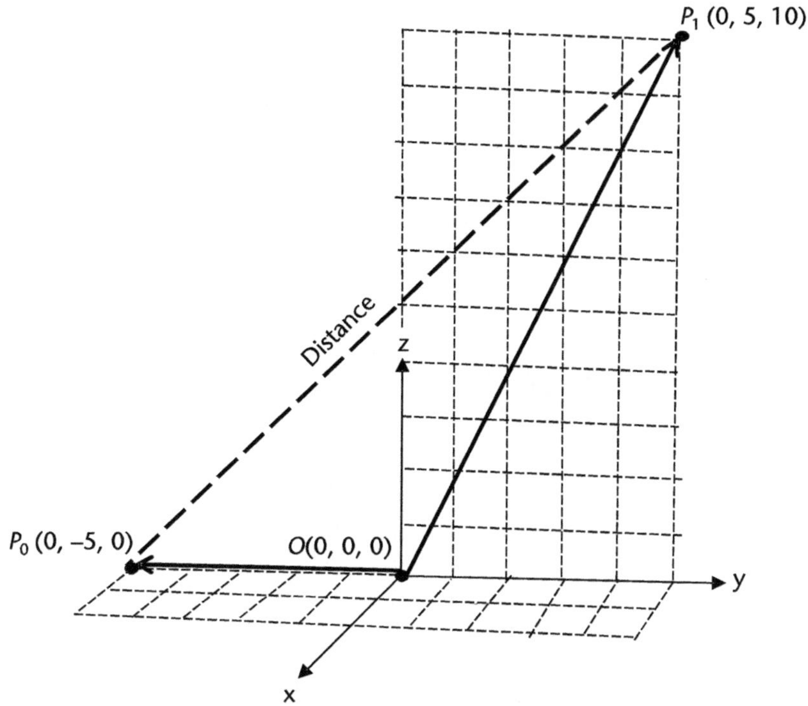

Figure 1.9 Vector diagram illustration for Problem 1.3.

Problem 1.4

See Figure 1.10. Given $\mathbf{A} = 2x + 4y - 3z$ and $\mathbf{B} = x - y$, find:

(a) $\mathbf{A} \cdot \mathbf{B}$,

(b) $\mathbf{A} \times \mathbf{B}$.

Problem 1.5

Show that:

$$\mathbf{A} = 4\mathbf{x} - 2\mathbf{y} - \mathbf{z} \quad \& \quad \mathbf{B} = \mathbf{x} + 4\mathbf{y} - 4\mathbf{z}$$

are perpendicular.

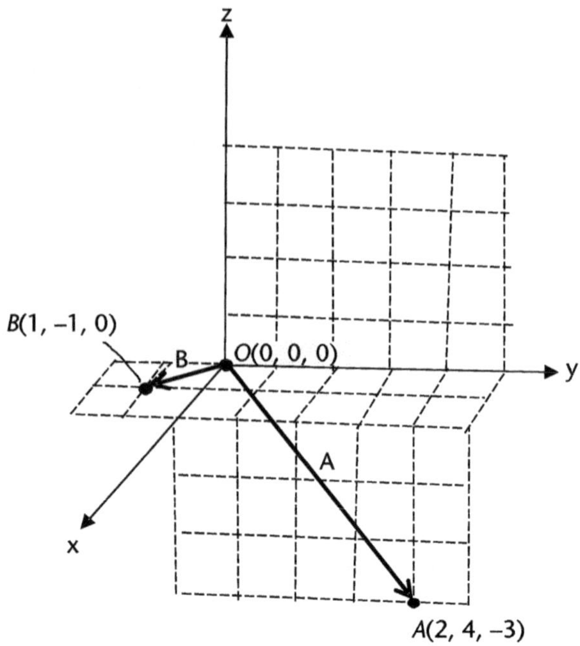

Figure 1.10 Vector diagram illustration for Problem 1.4.

Problem 1.6

Given that: $\mathbf{A} = 2\mathbf{x} + 4\mathbf{y}$ and $\mathbf{B} = 6\mathbf{y} - 4\mathbf{z}$, find the smaller angle between the vectors using the cross product and the dot product.

Problem 1.7

Given $\mathbf{F} = (y - 1)\mathbf{x} + (2x)\mathbf{y}$, find:

- The vector at (2, 2, 1),
- Its projection on B where $B = 5\mathbf{x} - \mathbf{y} + 2\mathbf{z}$.

Problem 1.8

Given:

$\mathbf{A} = \mathbf{x} + \mathbf{y}$,

$\mathbf{B} = \mathbf{x} + 2\mathbf{z}$,

$\mathbf{C} = 2\mathbf{y} + \mathbf{z}$,

find $(\mathbf{A} \times \mathbf{B}) \times \mathbf{C}$ and $\mathbf{A} \times (\mathbf{B} \times \mathbf{C})$.
Compare the answers.

Problem 1.9

Given:

$$\mathbf{A} = \mathbf{x} + \mathbf{y},$$
$$\mathbf{B} = \mathbf{x} + 2\mathbf{z},$$
$$\mathbf{C} = 2\mathbf{y} + \mathbf{z},$$

find $\mathbf{A} \cdot \mathbf{B} \times \mathbf{C}$, and find $\mathbf{A} \times \mathbf{B} \cdot \mathbf{C}$.
Compare the answers.

Problem 1.10

Express the unit vector that points from P_1 $(z_1 = h)$ on the z-axis toward $P_o(r, \phi, z_o)$, where $z_o = 0$, in cylindrical coordinates. See Figure 1.11.

Problem 1.11

Express the unit vector that is *directed toward the origin* from an arbitrary point on the plane, $z = -5$. Examine Figure 1.12.

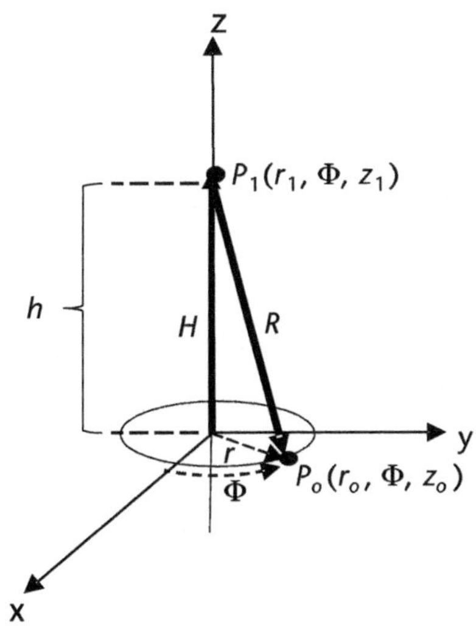

Figure 1.11 Vector diagram illustration for Problem 1.10.

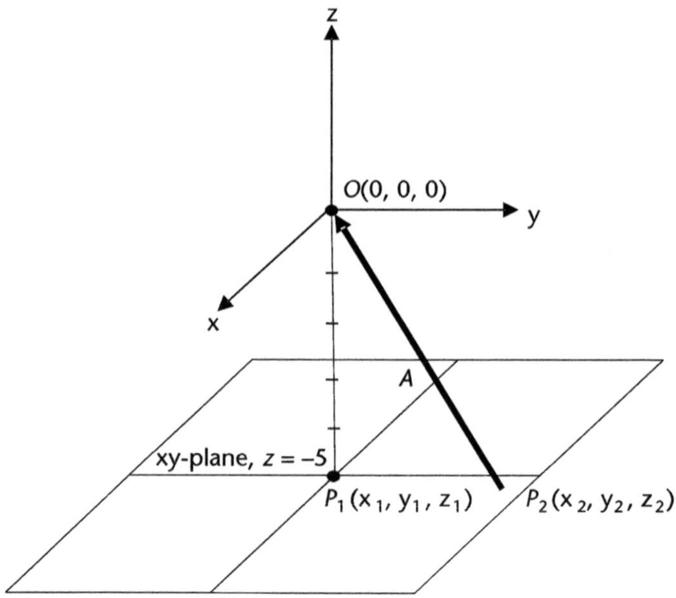

Figure 1.12 Vector diagram illustration for Problem 1.11.

Problem 1.12

Using the spherical coordinate system,

(a) Find the surface area of a strip $\alpha \le \Theta \le \beta$ on the spherical shell of radius α (see Figure 1.13).

(b) What is the result (area) when $\alpha = 0$ and $\beta = \pi$?

Problem 1.13

Develop the equation for the volume of a sphere of radius α from the differential volume equation.

Problem 1.14

Use the cylindrical coordinate system to find the area of the curved surface of a right circular cylinder where $r = 2\text{m}$, $h = 5$, and $30° < \phi < 120°$ (see Figure 1.14).

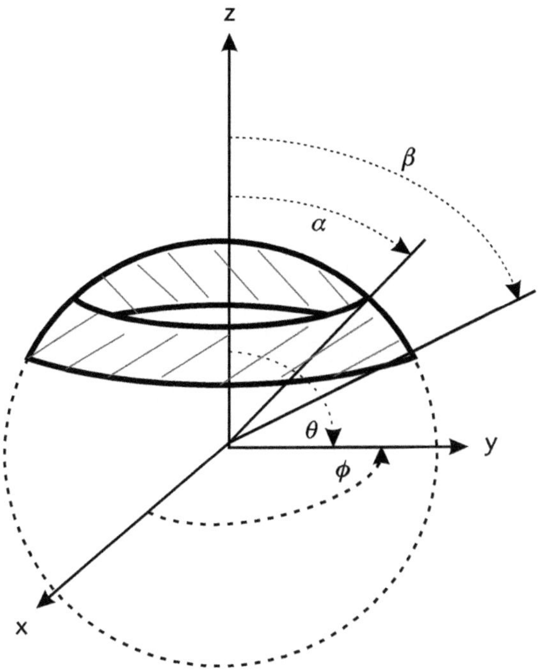

Figure 1.13 Vector diagram illustration for Problem 1.12.

Problem 1.15

Examine the cylinder of Figure 1.15. Quantify the surface areas of surfaces #1, #2, and #3 in the figure if the height of the cylinder is 10m, the inner cylinder radius, r_1, is 2m, and the outer cylinder radius, r_2, is 5m.

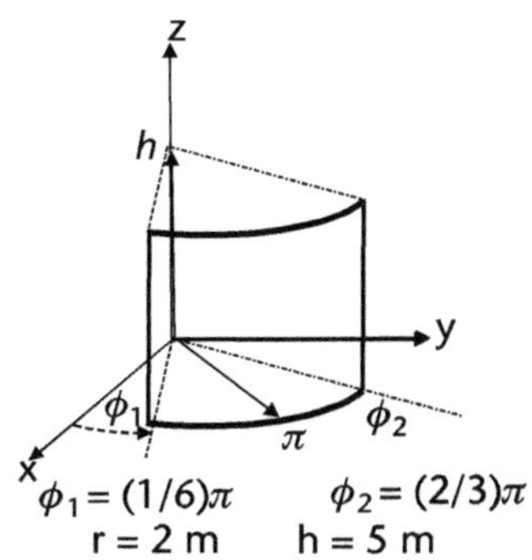

Figure 1.14 Vector diagram illustration for Problem 1.14.

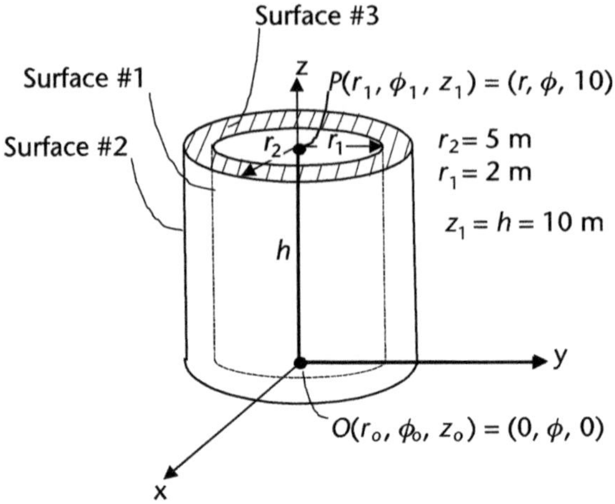

Figure 1.15 Vector diagram illustration for Problem 1.15.

References

[1] Boyce, W. E., *Calculus*, New York: John Wiley & Sons, 1988.

[2] Munein, M., and D. Foulis, *Calculus with Analytic Geometry*, New York: Worth Publishers, Inc., 1978.

[3] Anton, H., *Calculus with Analytic Geometry, Second Edition*, New York: John Wiley & Sons (Anton Textbooks, Inc.), 1984.

[4] Macnamara, T., *Handbook of Antennas for EMC*, Norwood, MA: Artech House, 1995.

[5] Stewart, J., *Calculus, Eighth Edition*, Boston, MA: Cengage Learning, 2016.

[6] Ayres, Jr., F., *Theory and Problems of Differential and Integral Calculus*, Second Edition, New York: McGraw-Hill, 1978.

2

COULOMB FORCES AND ELECTRIC FIELD INTENSITY

Stationery electrostatic charge generation occurs every day, and it is everywhere. It can be demonstrated to exist by way of:

- Walking inside a house on a room carpet in rubber-sole shoes on a dry-air wintry day and touching a metal doorknob;
- The pulling of Saran Wrap off the cylinder to cover a plastic salad dish (hence, the term "static cling");
- The removal of clothes from a warm dryer and noticing how, when you pull them apart, you see/hear arcs/sparks;
- The combing of dry hair in a dry winter climate using a rubber/plastic comb.

Other, less commonly known examples found in the real-world include static charge generation on the body and wings of a commercial jet (test) airplane occurring by flying above a rainstorm cloud at a (subzero temperature) altitude during a *p*-static flight test. During the test flight, the plane accumulates electrostatic (known as precipitation static) charge as evidenced by the need to bleed off the high charge remaining on the airplane after landing, parking, and using a large low-AWG conductive cable to bleed the accumulated charge off.

It may be of side interest for the reader to know that, of the four forces in the physical world, (i.e., nuclear, electrostatic, magnetostatic, and gravitational), electrostatic force is quite powerful and is the second most powerful force in the universe.

The fundamentals of electrostatics will be explored in this chapter along with introducing associated laws and equations. It should be noted that the focus of the next five chapters is on static charge, that is, charge that is not moving or changing in magnitude.

2.1 COULOMB'S LAW

Two electric charges, whether like or unlike, exert a force one on the other. If the charges are like-charges, the force is a *force of repulsion*. If the charges are unlike, the force is a *force of attraction*. Whichever the type, the force is directly proportional to the magnitude of each charge, Q_1 and Q_2, and the force is inversely proportional to the square of the separation distance between the charges. This is Coulomb's law. It was developed from direct observation of small, charged body interaction as made possible by use of a delicate torsion balance.

Throughout this text, calculations will be made in SI units. As such, units for repulsion or attraction force are in units of newtons (N). Units for the separation distance, d, between the charges are given in meters (m). The derived unit of charge is the coulomb (C). Force determination is rationalized by the product, 4π, and will be introduced via (2.3).

ε is the medium permittivity given with the units of C^2/Nm^2 or, equivalently (and, more commonly), F/m. Free-space permittivity (vacuum) is:

$$\varepsilon_0 = 8.854 \times 10^{-12} \quad \text{(F/m)} \tag{2.1}$$

For dielectric media other than free space, permittivity is a product of a material-specific dielectric constant, ε_r, and free-space permittivity, ε_0:

$$\varepsilon = \varepsilon_0 \varepsilon_r \quad \text{(F/m)} \tag{2.2}$$

where ε_r = *relative permittivity* or *dielectric constant of the media.*

In all solved problems throughout the text, it is important to know that free-space conditions are assumed unless otherwise so indicated. In vector form, Coulomb's law is stated as:

$$\mathbf{F} = \left(\frac{Q_1 Q_2}{4\pi\varepsilon d^2} \right) \mathbf{u}, \text{ (N)} \tag{2.3}$$

where

Q_1 = charge level of Pt 1 (C);

Q_2 = charge level of Pt 2 (C);

ε = dielectric material (or free space) permittivity between charges (F/m);

d = separation distance between charge points in space (m);

$\mathbf{u} = \mathbf{r}_{21}$ unit vector.

Subscripts on the unit vector (example, $\mathbf{r}_{21}$), help identify the force and indicate force direction:

$$\mathbf{F}_{21} = \left(\frac{Q_1 Q_2}{4\pi\varepsilon_0 d^2} \right) \mathbf{u}_{21} \text{ (N)} \tag{2.4}$$

Equation (2.4) describes the force, $\mathbf{F}_{21}$, *on* Q_1 *from* Q_2. Convention is that the notation $\mathbf{u}_{21}$ directs the vector from Q_2 to Q_1 (see Figure 2.1).

In the region around an isolated point charge, there is a spherically symmetrical force field that is made evident when a charge, Q, is fixed at the origin and charges of equal magnitude, Q_1 through Q_n, and like sign are positioned volumetrically around the origin (see Figure 2.2).

This can also be expressed in spherical coordinates given as follows:

$$\mathbf{F} = \left(\frac{Q_1 Q_2}{4\pi\varepsilon_0 r^2} \right) \mathbf{r} = \left(\frac{Q_1 Q_2}{4\pi\varepsilon_0 R^2} \right) \mathbf{r} \text{ (N)} \tag{2.5}$$

From a review of Figure 2.3, it can be seen that the symmetrical field about Q_o is disturbed by a second charge Q_1.

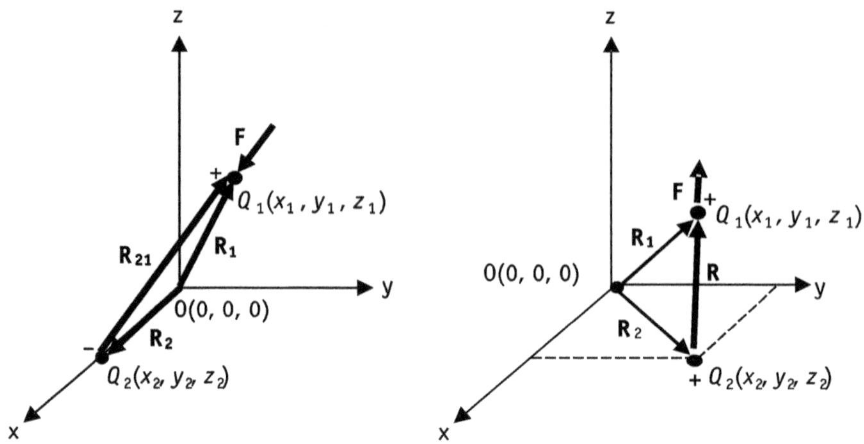

Figure 2.1 Two cases of a vector force, **F**, on a charge from another charge.

The resulting force, $\mathbf{F}_{result}$, is the vector sum of the two forces, $\mathbf{F}_0$ and $\mathbf{F}_1$, from charges Q_0 and Q_1,

$$\mathbf{F}_{result} = \mathbf{F}_0 + \mathbf{F}_1 \quad (\mathrm{N}) \tag{2.6}$$

Likewise, just as there is a spherical force field around Q_0, there is also a spherical force field that is around Q_1, too. The resulting field between the two fields is a point-by-point vector sum, that is, the *superposition principle* for coulomb forces, and extends to any number of charges [1].

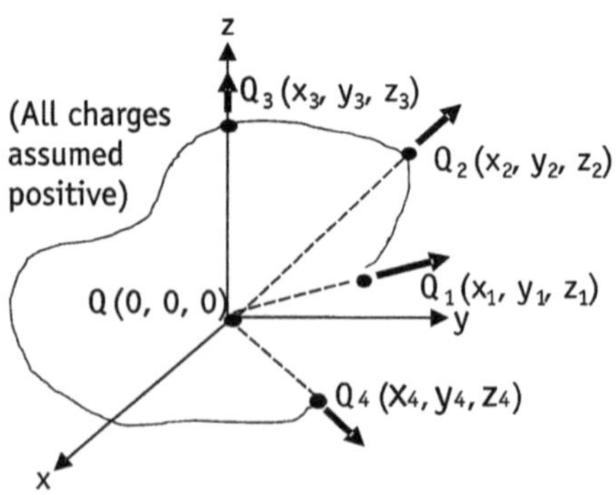

Figure 2.2 Illustration of force field around Q that is centered at the Cartesian coordinate system origin.

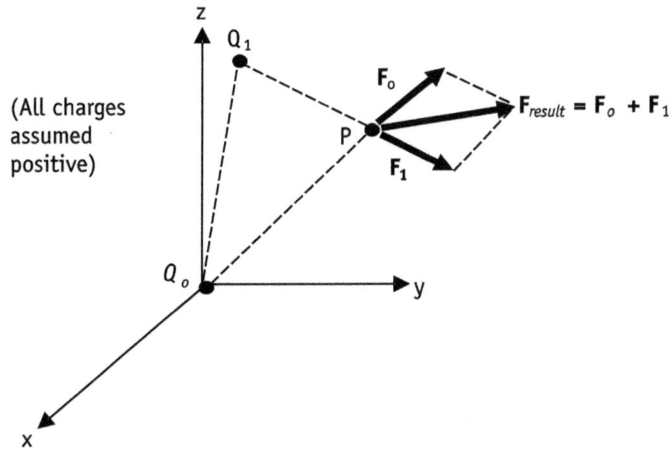

Figure 2.3 Illustration of electric forces together acting on a point in space, vector force sum result.

2.2 ELECTRIC FIELD INTENSITY

Suppose that in Figure 2.3, charge Q_1 is so small that it does not significantly disturb the fixed charge point Q_o (see Figure 2.4). In this case, the electric field (*E*-field) intensity, **E**, due to Q_o is defined to be the force per unit charge on Q_1, that is:

$$\mathbf{E} = \left(\frac{1}{Q_1}\right)\mathbf{F}_1 = \left(\frac{Q_0}{4\pi\varepsilon_0 R^2}\right)\mathbf{r}_1 \ \text{(V/m)} \tag{2.7}$$

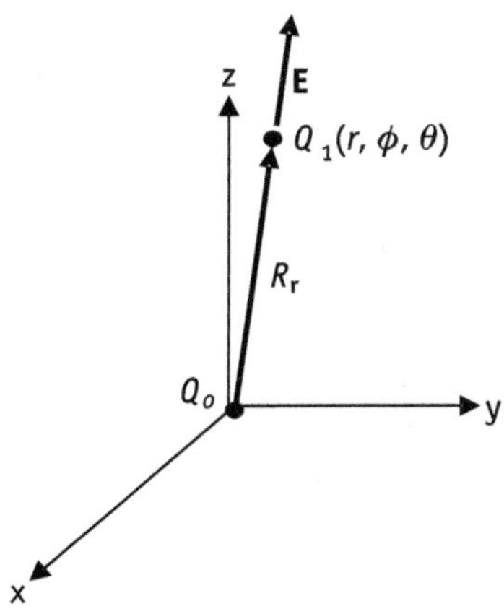

Figure 2.4 Illustration of electric field intensity and its vector (spherical coordinates).

In (2.7), **E** is given in spherical coordinates. The location of Q_o is placed at the origin. It may be transformed to other coordinate systems by the method described in Section 1.6.

In an arbitrary Cartesian coordinate system (see Figure 2.5), the equation for **E** may be expressed as follows:

$$\mathbf{E} = \left(\frac{Q}{4\pi\varepsilon_0 R^2}\right)\mathbf{R} \quad \text{(N/C or V/m)} \tag{2.8}$$

where the separation vector for R is **R**:

$$\mathbf{R} = (x_2 - x_1)\mathbf{x} + (y_2 - y_1)\mathbf{y} + (z_2 - z_1)\mathbf{z}.$$

The units for **E** are newtons per coulomb (N/C) as used by physicists. The more commonly used unit in the EM engineering field is volts per meter (V/m) [2].

2.3 CHARGE DISTRIBUTIONS

Up to this point, charge has been discussed in terms of a point or a set of points in space. There are other geometric distributions of charge that occur in the physical world.

Volume charge. When electric charge is distributed throughout a specified volume, each element of charge contributes to an electric field that can be seen at an external point. A volumetric integration

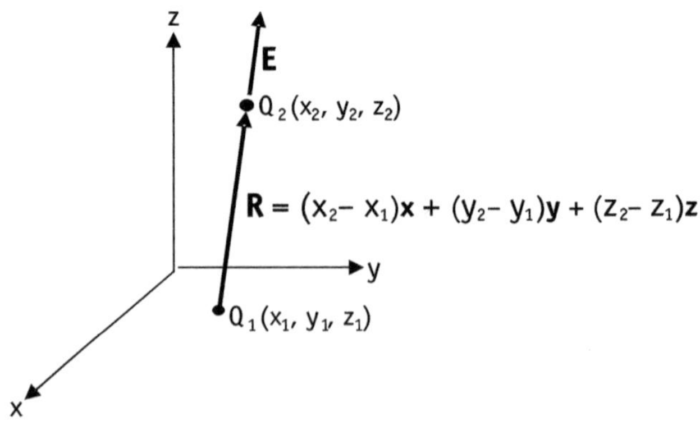

Figure 2.5 Illustration of electric field intensity vector (Cartesian coordinates).

is then required to obtain the total electric field. Even though electric charge in its smallest division is an electron or proton, it is useful to also consider charge as a continuous distribution, furthermore, one that is differentiable. As such, one may define and quantify the charge density by way of the following:

$$\rho_V = \frac{d}{dV}\,(Q) = \frac{dQ}{dV}\ \ (\mathrm{C/m^3}) \tag{2.9}$$

The unit in parentheses is meant to signify that ρV will be in $\mathrm{C/m^3}$, provided, of course, that the variables in the expression are quantified in proper SI units (i.e., C for charge, Q and $\mathrm{m^3}$ for volume, V). This convention is used throughout this book.

Referring to the Figure 2.6 volume, V, each differential charge, dQ, produces a differential electric field (E-field) that is quantified as at the observation point P.

$$d\mathbf{E} = \left(\frac{dQ}{4\pi\varepsilon_0 R^2}\right)\mathbf{r}\ \ (\mathrm{V/m}) \tag{2.10}$$

Under the assumption that the only charge in the region is contained within the volume, V, the total electric field at P is obtained by integration over the volume, that is:

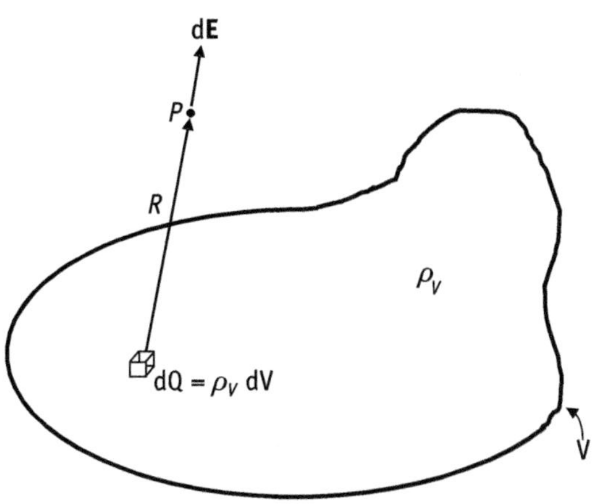

Figure 2.6 Generalized volumetric charge distribution.

$$\mathbf{E} = \int_V \left(\frac{\rho_V}{4\pi\varepsilon_0 R^2} \right) \mathbf{r}\, dV \quad \text{(V/m)} \tag{2.11}$$

A real-world application of a volumetric charge distribution is the case of commercial passenger or freighter airplane fuel tank being filled with Jet A-type kerosene. Before entering the fuel tank, the liquid dielectric (Jet A airplane fuel) passes through a hose that has an in-line particulate filter. While the filter removes impurities, that is, physical particulates, so that they do not enter the fuel tank, triboelectric friction created by the fuel passing across the filter elements generates electric charge. The generated electric charge flows down the remainder of the hose interior with the fuel to enter the tank through a designated opening. At the instant the fuel tank is filled with fuel, as indicated by the airplane fuel quantity sensors, like (positive) charge is distributed throughout the liquid fuel dielectric in a uniform charge distribution, making an easy-to-evaluate simplifying assumption.

Sheet charge. Charge may be also distributed over a surface or a sheet. Each differential charge, dQ, on the sheet results in a differential electric field,

$$d\mathbf{E} = \left(\frac{dQ}{4\pi\varepsilon_0 R^2} \right) \mathbf{r} \quad \text{(V/m)} \tag{2.12}$$

at point P (see Figure 2.7).

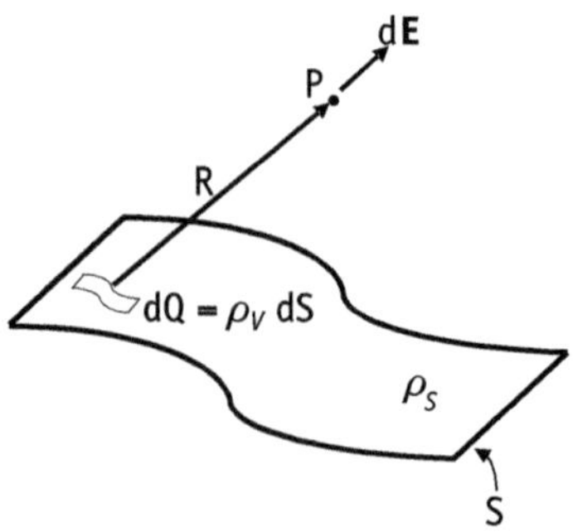

Figure 2.7 Generalized sheet (surface) charge distribution.

If the surface charge density is ρ_s (C/m^2) and no other charge is present in the region, then the total electric field at P is:

$$\mathbf{E} = \int_S \left(\frac{\rho_S}{4\pi\varepsilon_0 R^2} \right) \mathbf{r}\, dS \ \ \text{(V/m)} \tag{2.13}$$

Line charge. If charge is distributed over a line, each differential charge, dQ, along the sheet produces a differential electric field,

$$d\mathbf{E} = \left(\frac{dQ}{4\pi\varepsilon_0 R^2} \right) \mathbf{u}_R \ \ \text{(V/m)} \tag{2.14}$$

at point P (see Figure 2.8).

Furthermore, if the line charge density is ρ_l (C/m) and no other charge is in the region, the total electric field at P is,

$$\mathbf{E} = \int_L \left(\frac{\rho_l}{4\pi\varepsilon_0 R^2} \right) dl\ \mathbf{r} \ \ \text{(V/m)} \tag{2.15}$$

It should be emphasized that in all three of the above charge distributions and corresponding integrals for **E**, the unit variable, **r**, is variable depending on the coordinates of the charge element dQ. Thus, **r** cannot be removed from the integrand [1].

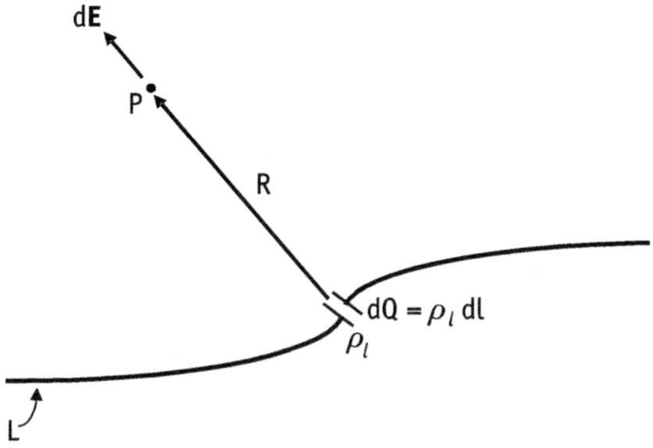

Figure 2.8 Generalized line charge distribution.

2.4 STANDARD CHARGE CONFIGURATIONS

In three special cases, the integration discussed in the previous section is either unnecessary or easily carried out. With regard to both these standard charge configurations (and to others that will be addressed later in the chapter), it should be pointed out that the charge is not on a conductor.

When a word problem states that charge is distributed in the form of a disk, for example, it does not mean that a disk-shaped conductor has charge on the surface. Although a stretch of the imagination may be required for this, these charges should be thought of as somehow suspended in space, fixed in the specified configuration.

Point charge. As determined in the previous section, the field of a single point charge Q is given by:

$$\mathbf{E} = \left(\frac{Q}{4\pi\varepsilon_0 R^2}\right)\mathbf{r} \text{ (spherical coordinates) (V/m)} \tag{2.16}$$

Figure 2.9 is a spherically symmetrical field that follows the inverse-square law (the same as for the force of gravitation).

Infinite line charge. If charge is distributed with uniform density, ρ_l (C/m) along an infinite straight line, which for convenience with be chosen as the z-axis (see Figure 2.10), then

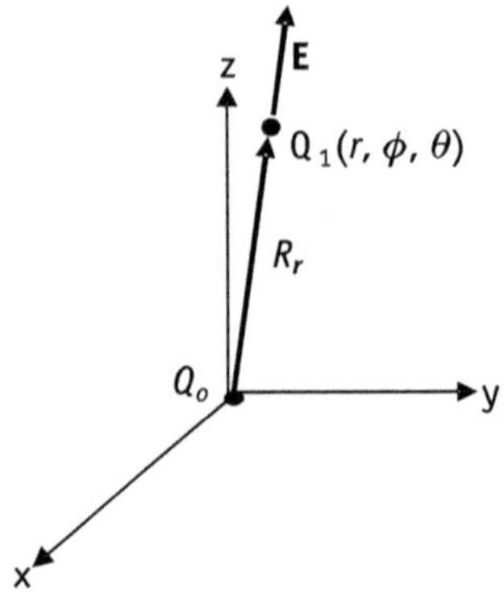

Figure 2.9 Spherically symmetric field (follows inverse-square law).

$$\mathbf{E} = \left(\frac{\rho_l}{2\pi\varepsilon_0 R} \right) \mathbf{r} \text{ (cylindrical coordinates) (V/m)} \tag{2.17}$$

This field has cylindrical symmetry and is inversely proportional to the first power of the distance from the infinite line charge.

Infinite plane charge. If charge is distributed with uniform density, ρ_S (C/m^2), over an infinite plane (see Figure 2.11), the field is then given by:

$$\mathbf{E} = \left(\frac{\rho_S}{2\varepsilon_0} \right) \mathbf{n} \tag{2.18}$$

This field has a constant magnitude and has mirror symmetry about the planar charge [2].

2.5 PROBLEMS: COULOMB FORCE AND E-FIELD INTENSITY

Problem 2.1

Find the force, $\mathbf{F}_1$, on Q_1 located at point (0, 1, 2) with charge of 20 μC due to charge Q_2 located at point (2, 0, 0) having charge of -300 μC.

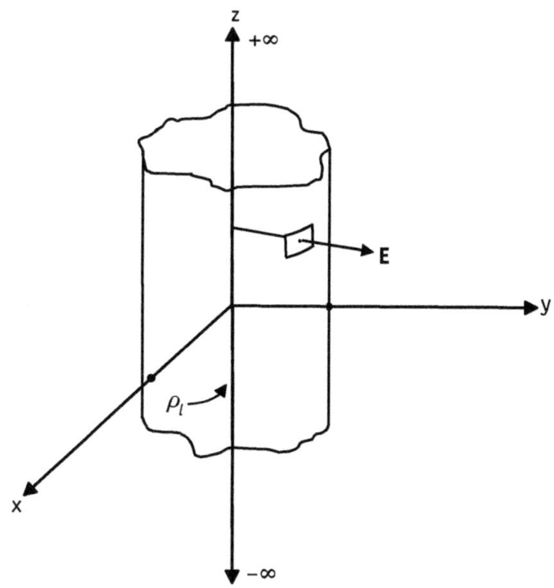

Figure 2.10 Infinite line charge, cylindrically symmetrical field.

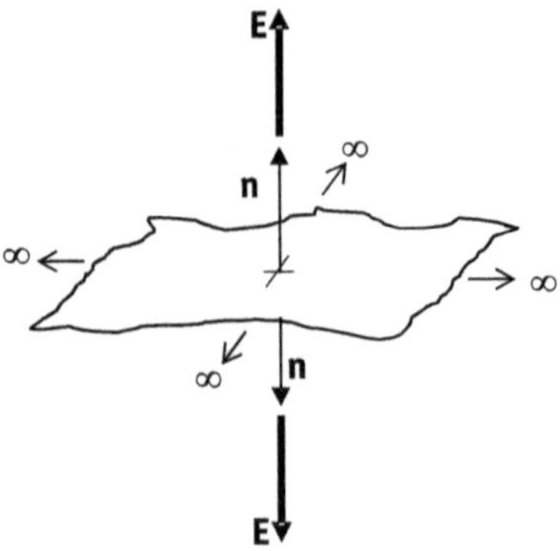

Figure 2.11 Infinite plane charge.

Problem 2.2

Two point charges, Q_1 and Q_2, are located at two separate points in space. Write a generalized equation to find the force of Q_2 on Q_1.

Problem 2.3

Two point charges, $Q_1 = 50\ \mu C$ and $Q_2 = 10\ \mu C$, are located, respectively, at (−1, 1, −3) and (3, 1, 0). Find the force on Q_1.

Problem 2.4

A point charge of $Q_1 = 300\ \mu C$ at (1, −1, −3)m experiences a force, $\mathbf{F} = 8\mathbf{x} - 8\mathbf{y} + 4\mathbf{z}$ N due to point charge Q_2 at (3, −3, −2). Determine Q_2.

Problem 2.5

Find the force on a point charge of 50 μC located at (0, 0, 5)m due a charge of $500\pi\ \mu C$ uniformly distributed over the circular disk $r \le 5$m, $z = 0$m.

Problem 2.6

Develop an expression for **E** due to charge uniformly distributed over an infinite plane with density, ρ_s.

Problem 2.7

Charge is distributed uniformly along an infinite straight line with density ρ_l. Develop the expression for **E** at the general point P.

Problem 2.8

Find the expression for **E** at the point P due to the point charge Q located at (x_1, y_1, z_1).

Problem 2.9

Find **E** at the point P (0, 0, 5) due to the point charge $Q_1 = 0.35\ \mu$C located at (0, 4, 0) and $Q_2 = -0.55\ \mu$C located at (3, 0, 0).

Problem 2.10

On the line described by $x = 2$m, $y = -4$m, that is, $(2, -4, z)$, there is a uniform charge distribution of density $\rho_l = 20$ nC/m. Describe the field, **E**, at location (−2, −1, 4).

Problem 2.11

Determine **E** at (2, 0, 2)m due to three standard charge distributions as follows:

- A uniform sheet a $x = 0$m with $\rho_{S1} = (1/3\pi)$ nC/m^2;
- A uniform sheet a $x = 4$m with $\rho_{S2} = (-1/3\pi)$ nC/m^2;
- A uniform sheet a $x = 4$m, $y = 0$m with $\rho_l = -2$ nC/m.

Problem 2.12

Charge is distributed along the z-axis between $z = \pm 5$m with uniform density $\rho_l = 20\ \mu$C/m. Determine **E** at (2, 0, 0) in:

(a) Cartesian coordinates;

(b) Cylindrical coordinates.

Problem 2.13

Find the electric field intensity, **E**, at the point P$(0, \phi, h)$ in cylindrical coordinates due to the uniformly charged disk $r \leq a$, $z = 0$.

Problem 2.14

Charge lies on a circular disk $r \leq a$ and $z = 0$ that has a charge density, $\rho_S = \rho_o \sin^2\phi$. Determine **E** at $(0, \phi, h)$.

Problem 2.15

Charge lies on the circular disk $r \leq 4$m, $z = 0$ with uniform density $\rho_S = 10^{-4}/r$ (C/m^2). Determine **E** at $r = 0$, $z = 3$m.

Problem 2.16

Charge line in the $z = -3$m plane in the form of a square sheet as defined by

$$2\text{m} \leq x \leq -2\text{m}$$

and

$$2\text{m} \leq y \leq -2\text{m}$$

with charge density $\rho_S = 2(x^2 + y^2 + 9)^{3/2}$ (nC/m^2). Find **E** at the origin.

Problem 2.17

A charge of uniform density, $\rho_S = 0.3$ nC/m^2 covers the plane of $2x - 3y + z = 6$m. Find **E** on the side of the plane facing the origin.

Problem 2.18

Two point charges, $Q_1 = 250\ \mu$C and $Q_2 = -300\ \mu$C, are located at (5, 0, 0)m and (0, 0, −5)m, respectively. Find the force on Q_2.

Problem 2.19

Two point charges, $Q_1 = 30\ \mu$C and $Q_2 = -100\ \mu$C, are located at (2, 0, 5)m and (−1, 0, −2)m, respectively. Find the force on Q_1.

References

[1] Shadowitz, A., *The Electromagnetic Field*, New York: Dover Publications, Inc., 1975.

[2] Russer, P., *Electromagnetics, Microwave Circuit and Antenna Design for Communications Engineering*, Norwood, MA: Artech House, 2006.

[3] Warnick, K. F., and P. Russer, *Problem Solving in Electromagnetics, Microwave Circuit and Antenna Design for Communications Engineering*, Norwood, MA: Artech House, 2006.

[4] Cheng, D. K., *Fundamentals of Engineering Electromagnetics*, Reading, MA: Addison-Wesley Publishing Company, Inc., 1993.

3

ELECTRIC FLUX AND GAUSS' LAW

In the previous chapter, Coulomb's law was introduced to the reader. Along with it were the electrostatic forces of repulsion (between like-charges) and attraction (between unlike charges). Also introduced was the concept of electric field intensity, and with it the concepts of volume charge, infinite sheet charge, infinite line charge, and point charge.

In this chapter, determination of net charge in a volume will be described along with the concepts of electric flux and flux density. The origin and termination points of these are illustrated and described. Gauss' law is introduced along with the relation between electric flux and flux density. Finally, the important and especially helpful special Gaussian surfaces (i.e., the imaginary Gaussian spherical shell (sphere) and special Gaussian cylinder are introduced).

3.1 NET CHARGE IN A REGION

With charge density defined in Section 2.3, it is at this point possible to quantify the net charge contained in a specified volume by way of integration. From:

$$dQ = \rho_V dV \quad \text{(C)} \tag{3.1}$$

it follows that:

$$Q = \int_V \rho_V \, dV \quad \text{(C)} \tag{3.2}$$

Of course, ρ_V need not be constant throughout the volume, V [1].

3.2 ELECTRIC FLUX AND FLUX DENSITY

By definition and convention, electric flux, Ψ, originates on positive charge and terminates on negative charge. In the absence of a terminating negative charge, the flux, Ψ, terminates at infinity.

Also, by definition, one coulomb (C) of electric charge gives rise to one coulomb of electric flux. Hence:

$$\psi = Q \quad \text{(C)} \tag{3.3}$$

In Figure 3.1, the flux lines leave $+Q$ and terminate on $-Q$. The assumption in the figure is that the two unlike charges are of equal magnitude.

The case of the positive charge with no negative charge to terminate the flux, as shown in Figure 3.2, illustrates that flux lines are equally spaced throughout the solid angle and reach out to infinity.

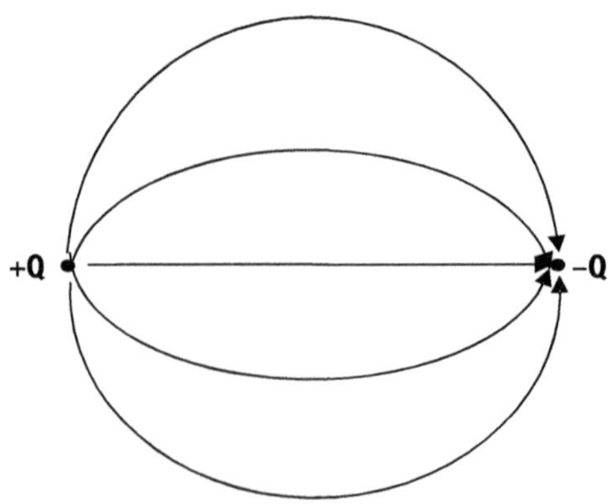

Figure 3.1 Origination and termination of electric flux, point charge to point charge.

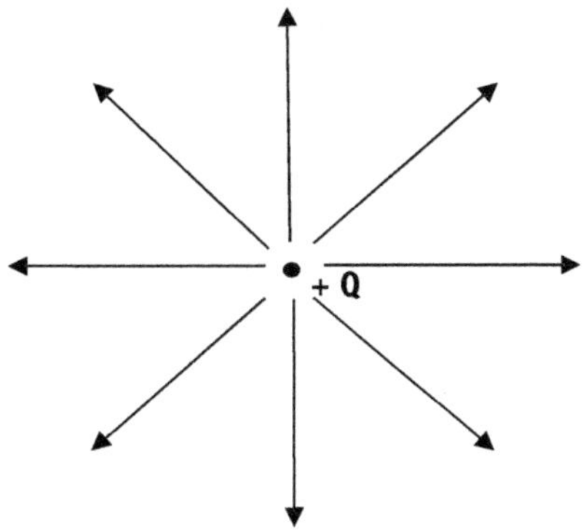

Figure 3.2 Origination and termination of electric flux point to infinity in all directions.

While the electric flux, Ψ, is a scalar quantity, the *density of electric flux*, **D**, is a vector field. As such, it takes its direction from the lines of flux. If in the neighborhood of point P, the flux lines have the direction of the unit vector, **a**, as illustrated in Figure 3.3.

Furthermore, if an amount of flux, $d\psi$, crosses the differential area, dS, which is normal to **a**, then the electric flux density at P is [2]:

$$\mathbf{D} = \frac{d\Psi}{dS}\mathbf{a} \quad \left(\mathrm{C/m^2}\right) \tag{3.4}$$

A volume charge distribution of density, ρ_V (C/m³), is shown enclosed by surface, S, in Figure 3.4.

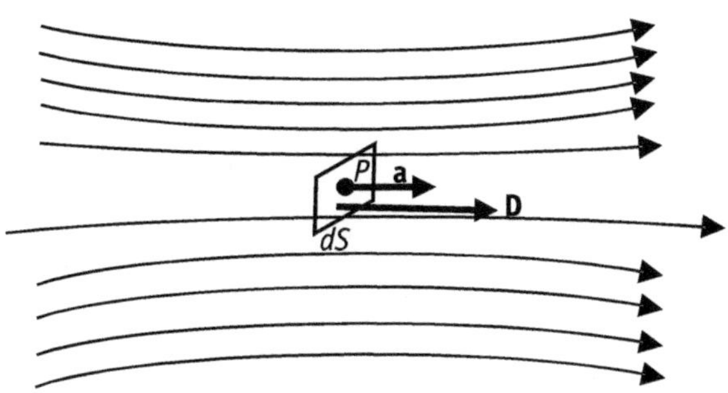

Figure 3.3 Lines of flux through point P and an elemental surface area, *dS*.

Since each coulomb of charge Q has, by definition, 1C of flux, Ψ, it follows that the net flux crossing the closed surface, S, is an exact measure of net charge enclosed. This said, the density, **D**, may vary in magnitude and direction from point to point on S. So, generally speaking, **D** will not be along the normal line to S. If at the surface of element dS, **D** makes an angle θ with the normal line, then the differential flux crossing dS is given by

$$d\Psi = D\,dS\cos\theta = \mathbf{D}\cdot dS\,\mathbf{a}_n = \mathbf{D}\cdot d\mathbf{S} \tag{3.5}$$

where $d\mathbf{S}$ = the vector surface element of magnitude dS and direction $\mathbf{a}_n$.

The unit vector, $\mathbf{a}_n$ (a.k.a., $\mathbf{u}_n$ or $\mathbf{n}$) is always taken to point out of S so that $d\Psi$ is the amount of flux passing from the interior of S to the exterior of S through dS [3].

3.3 GAUSS' LAW

Integration of the above expression for $d\Psi$ over the closed surface S gives since $\Psi = Q$,

$$\oint \mathbf{D}\cdot d\mathbf{S} = Q_{\text{enc}} \text{ (Gauss' Law)} \tag{3.6}$$

The above equation is Gauss' law. It states that the total flux out of a closed surface is equal to the net charge within that surface.

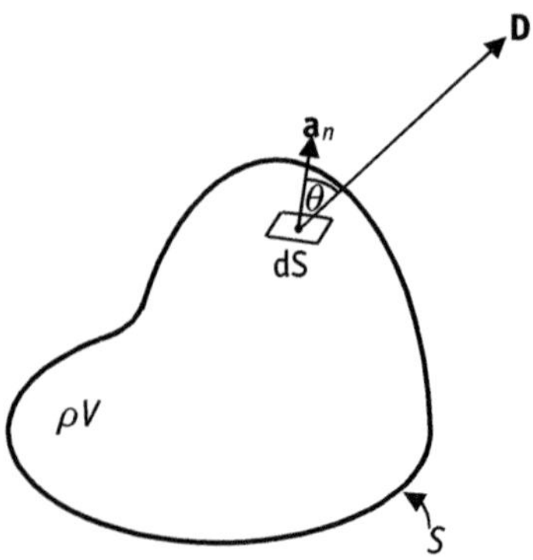

Figure 3.4 A volume charge density, ρ_V (C/m^3) with a surface area enclosing the volume.

It will be seen that a great deal of valuable information can be obtained from applying Gauss' law without having to perform the integration [2].

3.4 RELATION BETWEEN FLUX DENSITY AND ELECTRIC FIELD INTENSITY

Consider a point charge Q—assumed to be positive for simplicity—that is located at the origin (see Figure 3.5).

If the point charge is enclosed by a spherical surface having a radius, r, then by symmetry, $\mathbf{D}$, due to Q, is of constant magnitude over the surface. What's more, it is *everywhere* normal to the surface. Gauss' law then gives:

$$Q = \oint \mathbf{D} \cdot d\mathbf{S} = D\oint dS = D\left(4\pi r^2\right) \tag{3.7}$$

from which $D = Q/4\pi r^2$. For this reason,

$$\mathbf{D} = \frac{Q}{4\pi r^2}\mathbf{a}_n = \frac{Q}{4\pi r^2}\mathbf{a}_r = \frac{Q}{4\pi r^2}\mathbf{n} = \frac{Q}{4\pi r^2}\mathbf{r} \tag{3.8}$$

However, from Section 2.2, the electric field intensity due to Q is:

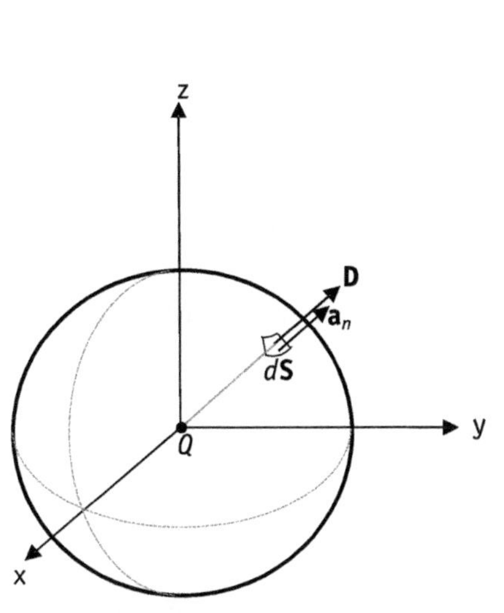

Figure 3.5 Point charge, Q, enclosed by a sphere (Gauss' law illustration).

$$\mathbf{E} = \frac{Q}{4\pi\varepsilon_0 r^2}\mathbf{a}_r = \frac{Q}{4\pi\varepsilon_0 r^2}\mathbf{r} \tag{3.9}$$

It follows that:

$$\mathbf{D} = \varepsilon_0 \mathbf{E} \tag{3.10}$$

More generally speaking, any electric field in an isotropic medium of permittivity, ε, will have

$$\mathbf{D} = \varepsilon \mathbf{E} \tag{3.11}$$

Thus, **D** and **E** fields will have identically the same form. They differ only by a factor which is a constant of the dielectric medium involved. The *e*-field, **E**, due to a charge configuration is a function of the permittivity ε. However, the electric flux density **D** is not. In situations involving multiple dielectric materials, a distinct advantage will be found in first obtaining **D** and then later converting to **E** for each dielectric material [2, 3].

3.5 SPECIAL GAUSSIAN SURFACES

The spherical surface used in Section 3.4 was a special Gaussian surface. That Gaussian surface has, and satisfies, the following defining conditions:

1. The surface is closed;
2. **D** is either normal or tangential to the surface at each point of the surface;
3. D has the same value at all points of the surface where **D** is normal.

It is a special Gaussian surface known as the spherical surface (see Figure 3.6).

A second type of special Gaussian surface is the cylindrical surface (see Figure 3.7).

One serious limitation to the method of special Gaussian surfaces is they can only be utilized for *highly symmetrical* charge

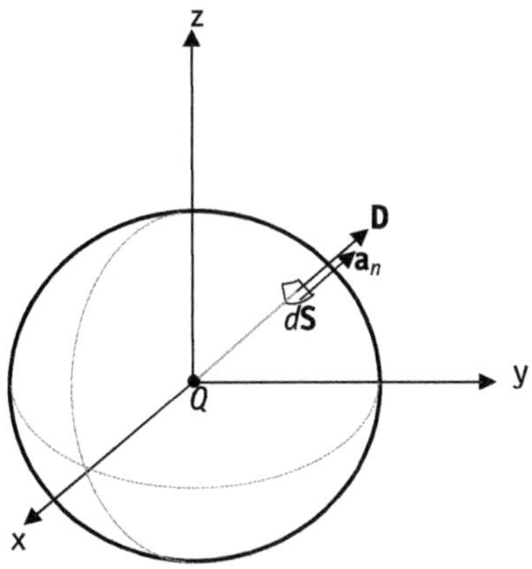

Figure 3.6 Special Gaussian surface: spherical surface.

configurations. For other configurations, the method can yet provide quick approximations to the field at locations very close to or very far from the charges [1].

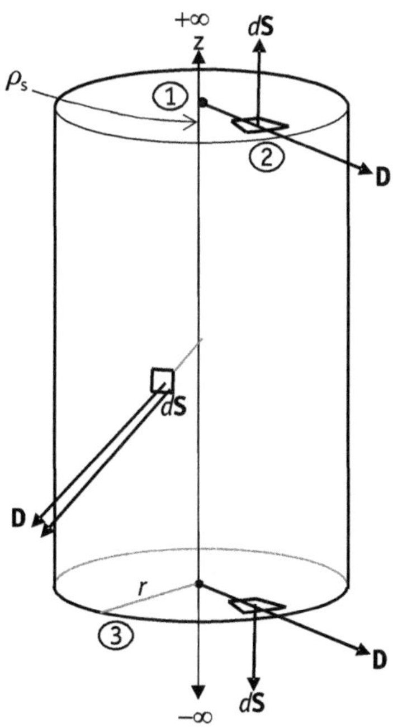

Figure 3.7 Special Gaussian surface: cylindrical surface.

3.6 PROBLEMS AND SOLUTIONS: ELECTRIC FLUX AND GAUSS' LAW

It is important to note the following:

1. **D** and **E** fields will have exactly the same form, differing only by ε, constant of the medium.
2. In problems involving multiple dielectrics, a distinct advantage will be found in first obtaining **D** and then converting to **E** within each dielectric.
3. The special Gaussian surfaces are conceptual geometric shapes (i.e., the imaginary spherical shell and the cylindrical shell), that help to more easily find solutions to electrostatics problems [3].

Problem 3.1

Use the special Gaussian surface to find D due to a uniform line charge, ρ_l (C/m).

Problem 3.2

Use the special Gaussian surface, the imaginary sphere shell, to find **D** due to point charge, Q (C), located at the sphere's center.

Problem 3.3

Find **E** in Problem 3.2.

Problem 3.4

Find **E** in Problem 3.1.

Problem 3.5

Find the charge in the volume shown in Figure 3.6 defined by:

(a) $0 \le x \le 1\text{m}$
$0 \le y \le 1\text{m}$
$0 \le z \le 1\text{m}$

and

(b) $0 \le x \le 1\text{m}$
$-1 \le y \le 0$
$0 \le z \le 1\text{m}$

if $\rho_V = 30\, x^2 y\ (\mu\text{C/m}^3)$.

Problem 3.6

Find the charge in the volume defined by $1 \leq r \leq 2$m in spherical coordinates if $\rho_V = 5\cos^2\phi/r^4$ (C/m^3).

Problem 3.7

Three point charges (i.e., $Q_1 = 30$ nC, $Q_2 = 150$ nC, and $Q_3 = -70$ nC), are enclosed by a surface, S. What net flux crosses S?

Problem 3.8

What net flux crosses the closed surface, S, which contains a charge distribution in the form of a plane disk of radius 4m having a density of $\rho_S = \sin\phi$ (μC/m^2)?

Problem 3.9

A circular disk of radius 4m having a surface charge density of $\rho_S = 12\sin\phi \times 10^{-6}$ C/m^2 is enclosed by surface S. What net flux crosses S?

Problem 3.10

Two charges of the same magnitude but opposite signs are enclosed by surface, S. Can flux Ψ cross the surface?

Problem 3.11

Charge in the form of a plane sheet with density $\rho_S = 40\ \mu$C/m^2 is located at $z = 0.5$m. A uniform charge line of $\rho_l = -6\ \mu$C/m lies along the y-axis. What net flux crosses the surface of a cube 2m on an edge centered at the origin?

Problem 3.12

A point charge Q is at the origin of a spherical coordinate system.

(a) Find the flux that crosses the portion of a spherical shell described by $\alpha \leq \phi \leq \beta$.
(b) What is the result if $\alpha = 0$ and $\beta = \pi/2$?

Problem 3.13

A uniform line charge with $\rho_l = 50\ \mu$C/m lies along the x-axis.

What flux per unit length, Ψ/L, crosses the portion of the $z = -3$m plane bounded by $y = \pm 2$m?

Problem 3.14

A point charge, $Q = 30$ nC, is located at the origin in Cartesian coordinates. Find the electric flux density, **D**, at (1, 3, −4)m.

Problem 3.15

Two identical uniform line charges lie along the x- and y-axes with charge densities, $\rho_l = 20\ \mu$C/m. Obtain **D** at (3, 3, 3)m.

Problem 3.16

Determine the flux crossing an area of 1 square millimeter on the surface of a cylinder shell at:

$R = 10$m, $\Phi = 53.2°$, and $Z = 2$m

if the surface area flux density (C/m^2) is:

$\mathbf{D} = (2x)\mathbf{x} + 2(1 - y)\mathbf{y} + (4z)\mathbf{z}$ (C/m^2).

Problem 3.17

Given an electric flux density, $\mathbf{D} = (2x)\mathbf{x} + 3\mathbf{y}$ (C/m^2), determine the net flux crossing the surface of a cube 2m on an edge centered at the origin (cube edges are parallel to the Cartesian coordinate axes).

Problem 3.18

A uniform line charge of $\rho_l = 3\ \mu$C/m lies along the z-axis, and a concentric cylinder of $r = 2$m has a surface charge distribution of $\rho_S = -(1.5/4\pi)\ \mu$C/m^2. Both distributions are infinite in extent with the z-axis. Use Gauss' law to find **D** in region.

Problem 3.19

A charge configuration in cylindrical coordinates is given by $\rho_V = 5re^{-2r}$ (C/m^3).

References

[1] Shadowitz, A., *The Electromagnetic Field*, New York: Dover Publications, Inc., 1975.

[2] Cheng, D. K., *Fundamentals of Engineering Electromagnetics*, Reading, MA: Addison-Wesley Publishing Company, Inc., 1993.

[3] Russer, P., *Electromagnetics, Microwave Circuit and Antenna Design for Communications Engineering*, Norwood, MA: Artech House, 2006.

[4] Warnick, K. F., and P. Russer, *Problem Solving in Electromagnetics, Microwave Circuit and Antenna Design for Communications Engineering*, Norwood, MA: Artech House, 2006.

4

DIVERGENCE AND THE DIVERGENCE THEOREM

In this chapter, an introductory discussion on divergence is undertaken, and it is undertaken in the three coordinate system forms: (1) Cartesian, (2) cylindrical, and (3) spherical. The del operator is introduced as is divergence theorem. Problems and solutions involving all the above are given at the end of the chapter.

4.1 DIVERGENCE

There are two characteristics regarding the manner in which a vector field changes from point to point in space. The first of these characteristics is *divergence*. Divergence is a *scalar* quantity, bearing similarity to the derivative of a function. The second characteristic is *curl*, a vector to be examined during the discussion of magnetic fields in Volume 2 [1].

When the divergence of a vector field is nonzero, the region is said to contain *sources* or *sinks*. Sources occur when the divergence is positive. Sinks occur when the divergence is negative.

In static electric fields, a correspondence exists between positive divergence, sources, and positive electric charge Q. Electric flux, Ψ, by definition, originates on positive charge. Therefore, a region that contains positive charges also contains the sources of Ψ. Divergence of the electric flux density $\mathbf{D}$ will be positive in this region. A similar correspondence exists as well between negative divergence, sinks, and negative electric charge.

Divergence of the vector field **A** at the point P is defined by:

$$\text{div}A = \lim_{\Delta V \to 0} \oint \mathbf{A} \cdot \frac{d\mathbf{S}}{\Delta V} \tag{4.1}$$

where V = volume.

In this case, the integration is over the surface of an infinitesimal volume, ΔV, which shrinks to point P.

4.2 DIVERGENCE IN CARTESIAN COORDINATES

The divergence can be expressed for any vector field and in any coordinate system. For development in Cartesian coordinates, a cube is selected with edges Δx, Δy, and Δz parallel to the x-, y-, and z-axes as illustrated in Figure 4.1.

The vector field, **A**,

$$\mathbf{A} = A_x\mathbf{x} + A_y\mathbf{y} + A_z\mathbf{z} \tag{4.2}$$

is then defined at point P; that is, at the cube corner with the lowest values of x, y, and z coordinates.

In order to express $\oint \mathbf{A} \cdot d\mathbf{S}$ for the cube, all six faces must be addressed in the analysis. On each face, the direction of $d\mathbf{S}$ is out-

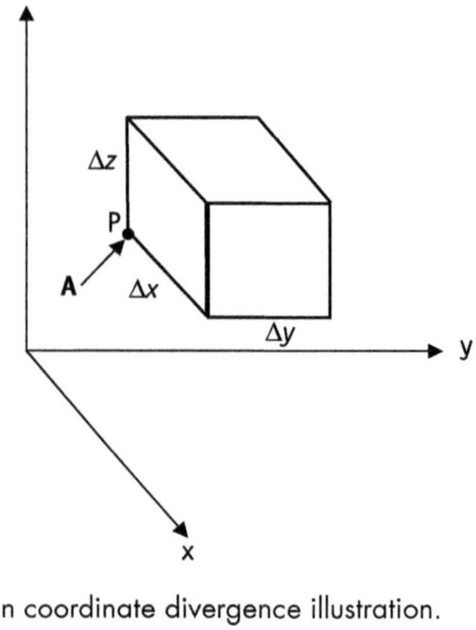

Figure 4.1 Cartesian coordinate divergence illustration.

ward. Since the faces are normal to the three coordinate axes, only one component of **A** crosses any two parallel faces (see Figure 4.2).

In Figure 4.2, the cube is turned such that face 1 is in full view. The x-components of A that are over the faces to the left and right of face 1 are indicated. Since the faces are small [2],

Left face:

$$\int \mathbf{A} \cdot d\mathbf{S} \sim -A_x(x)\Delta y \Delta z \tag{4.3}$$

Right face:

$$\int \mathbf{A} \cdot d\mathbf{S} \sim A_x(x + \Delta x)\Delta y\, \Delta z = \left[A_x(x) + \left(\frac{\partial A_x}{\partial x} \right) \Delta x \right] \Delta y\, \Delta z \tag{4.4}$$

the total for these two faces is:

$$\left(\frac{\partial A_x}{\partial x} \right) \Delta x\, \Delta y\, \Delta z \tag{4.5}$$

The same procedure is applied to the remaining two pairs of faces and the results combined.

$$\int \mathbf{A} \cdot d\mathbf{S} = \left(\frac{\partial A_x}{\partial x} + \frac{\partial A_y}{\partial y} + \frac{\partial A_z}{\partial z} \right) \Delta x\, \Delta y\, \Delta z \tag{4.6}$$

Dividing by

$$\Delta x\, \Delta y\, \Delta z = \Delta V \tag{4.7}$$

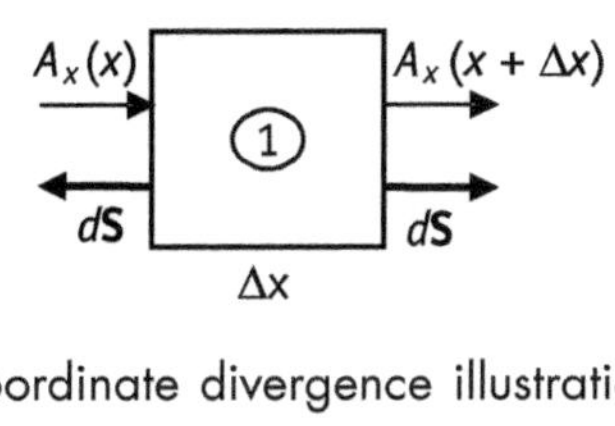

Figure 4.2 Cartesian coordinate divergence illustration, cube turned, face in full view.

and letting $\Delta V \to 0$, one obtains:

$$\operatorname{div}\mathbf{A} = \left(\frac{\partial A_x}{\partial x} + \frac{\partial A_y}{\partial y} + \frac{\partial A_z}{\partial z}\right) \text{ (expression in Cartesian coordinates)} \tag{4.8}$$

The same approach may be used in the use of cylindrical and spherical coordinates:

Cylindrical coordinates:

$$\operatorname{div}\mathbf{A} = \left[\left(\frac{1}{r}\right)\frac{\partial}{\partial r}\left(r A_r\right) + \left(\frac{1}{r}\right)\frac{\partial A_\phi}{\partial \phi} + \frac{\partial A_z}{\partial z}\right) \tag{4.9}$$

Spherical coordinates:

$$\operatorname{div}\mathbf{A} = \left\{\left(\frac{1}{r^2}\right)\frac{\partial}{\partial r}\left(r^2 A_r\right) = \frac{1}{(r\sin\theta)}\left[\frac{\partial}{\partial \theta}\left(A_\theta \sin\theta\right)\right] + \left(\frac{1}{\sin\theta}\right)\left(\frac{\partial A_\phi}{\partial \phi}\right)\right\} \tag{4.10}$$

4.3 DIVERGENCE OF *D*

From Gauss' law (as described in Section 3.3),

$$\frac{\left(\oint \mathbf{D}\cdot d\mathbf{S}\right)}{\Delta V} = \frac{Q_{\text{enc}}}{\Delta V} \tag{4.11}$$

In the limit,

$$\lim_{\Delta V \to 0} \oint \mathbf{D}\cdot\frac{dS}{\Delta V} = \operatorname{div}\mathbf{D} = \lim_{\Delta V \to 0}\left(\frac{Q_{\text{enc}}}{\Delta V}\right) = \rho \tag{4.12}$$

This important result is one of James Clerk Maxwell's equations for static fields:

$$\operatorname{div}\mathbf{D} = \rho \tag{4.13}$$

and

$$\operatorname{div}\mathbf{E} = \frac{\rho}{\varepsilon} \tag{4.14}$$

If ε is constant throughout the region under examination (if not, div ε $\mathbf{E} = \rho$). Thus, both **E** and **D** fields will have divergence of zero in any charge-free region.

4.4 THE DEL OPERATOR

Vector analysis has its own shorthand, and, as such, the reader must note it with care. At this point in the text, a vector operator, symbolized by ∇ is defined—in Cartesian coordinates—by:

$$\nabla = \frac{\partial(\;)}{\partial x}\mathbf{x} + \frac{\partial(\;)}{\partial y}\mathbf{y} + \frac{\partial(\;)}{\partial z}\mathbf{z} \tag{4.15}$$

Please note that the "( )" indicates the place in (4.15) on which to perform the del operator. In calculus, a differential operator D is sometimes used to represent d/dx. Note that the symbol's square root radical and $\int$ are also operators. Standing alone without any indication of what they are to operate on, they look strange [3].

Likewise, ∇ standing alone, simply suggests the taking of certain partial derivatives, each being followed by a unit vector. However, when ∇ is dotted with vector **A**, the result is the divergence of **A**.

$$\begin{aligned}\nabla \cdot \mathbf{A} &= \left(\frac{\partial}{\partial x}\mathbf{a}_x + \frac{\partial}{\partial y}\mathbf{a}_y + \frac{\partial}{\partial z}\mathbf{a}_z\right)\cdot\left(A_x\mathbf{a}_x + A_y\mathbf{a}_y + A_z\mathbf{a}_z\right)\\ &= \frac{\partial}{\partial x}\left(A_x\right) + \frac{\partial}{\partial y}\left(A_y\right) + \frac{\partial}{\partial z}\left(A_z\right) = div\,\mathbf{A}\end{aligned} \tag{4.16}$$

From here and forward, the divergence of a vector field, **A**, will be written as $\nabla \cdot \mathbf{A}$.

WARNING! *The del operator is defined ONLY in Cartesian coordinates.* When $\nabla \cdot \mathbf{A}$ is written for the divergence in other coordinate systems, it does not mean that a del operator can be defined for these systems. For example, the divergence in cylindrical coordinates will be written as:

$$\nabla \cdot \mathbf{A} = \left(\frac{1}{r}\right)\frac{\partial}{\partial r}(rA_r) + \left(\frac{1}{r}\right)\frac{\partial}{\partial \phi}(A_\phi) + \frac{\partial}{\partial z}(A_z) \quad \text{(See Section 4.2)} \tag{4.17}$$

It should be noted that *this does not imply* that:

$$\nabla = \left(\frac{1}{r}\right)\frac{\partial}{\partial r}(\ )\mathbf{a}_r + \left(\frac{1}{r}\right)\frac{\partial}{\partial \phi}(\ )\mathbf{a}_\phi + \frac{\partial}{\partial z}(\ )\mathbf{a}_z \tag{4.18}$$

in cylindrical coordinates. In fact, the expression would give false results when used in ∇V; that is, the gradient (see Chapter 5) or $\nabla \times \mathbf{A}$ (i.e., the curl).

4.5 DIVERGENCE THEOREM

Gauss' law states that the enclosed surface integral of $\mathbf{D} \cdot d\mathbf{S}$ is equal to the enclosed charge if the charge density function, ρ_V, is known throughout the volume [4]. Thus,

$$\oint \mathbf{D} \cdot d\mathbf{S} = \int \rho_V \, dV = Q_{\text{enc}} \tag{4.19}$$

But,

$$\rho_V = \nabla \cdot \mathbf{D} \tag{4.20}$$

So,

$$\oint \mathbf{D} \cdot d\mathbf{S} = \int [\nabla \cdot \mathbf{D}] dV \tag{4.21}$$

This is the *divergence theorem* (also known as *Gauss' divergence theorem*); that is, a three-dimensional analog of Green's theorem for the plane. While it was derived from known relationships among **D**, Q, and ρ_V, the theorem is applicable to *any* vector field.

$$\text{Divergence theorem} \quad \oint_S \mathbf{A} \cdot d\mathbf{S} = \int_V [\nabla \cdot \mathbf{A}]\, dV \tag{4.22}$$

Of course, the volume, V, is that which is enclosed by the surface, S.

The divergence theorem applies to time-variation fields as well as static fields in any coordinate system. The theorem is used most often in derivations where it becomes necessary to change from a closed surface integration to a volume integration. However, it may also be used in the reverse direction, too; that is, to convert the volume integral of a function that can be expressed as the divergence of a vector field into a closed surface integral.

4.6 PROBLEMS AND SOLUTIONS: DIVERGENCE AND THE DIVERGENCE THEOREM

Problem 4.1

Develop the expression for divergence in cylindrical coordinates (Figure 4.3).

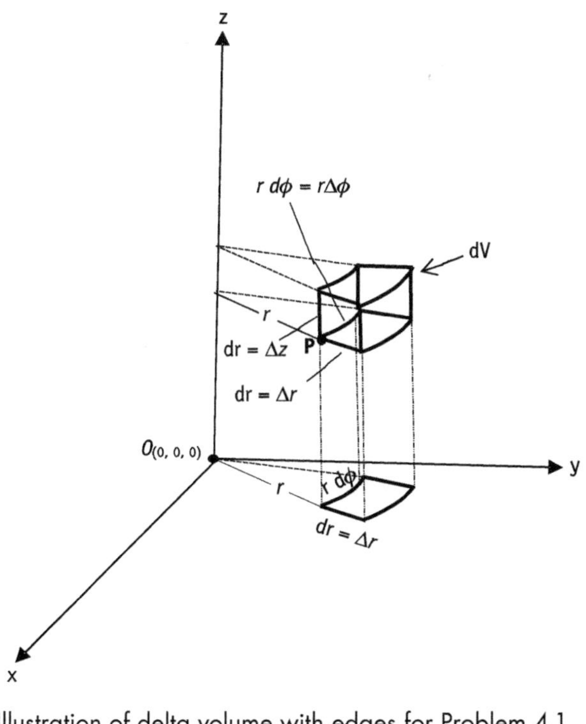

Figure 4.3 Illustration of delta volume with edges for Problem 4.1.

Problem 4.2

Show that $\nabla \cdot \mathbf{E} = 0$ for the field of a uniform line charge.

Problem 4.3

Show that the D field due to a point charge has a divergence of zero.

Problem 4.4

Given $\mathbf{A} = e^{-y} [\cos x\mathbf{x} - \sin x\mathbf{y}]$, find $\nabla \cdot \mathbf{A}$.

Problem 4.5

Given $A = x^2\mathbf{x} + (yz)\mathbf{y} + (xy)\mathbf{z}$, find $\nabla \cdot \mathbf{A}$.

References

[1] Cheng, D. K., *Fundamentals of Engineering Electromagnetics*, Reading, MA: Addison-Wesley Publishing Company, Inc., 1993.

[2] Shadowitz, A., *The Electromagnetic Field*, New York: Dover Publications, Inc., 1975.

[3] Russer, P., *Electromagnetics, Microwave Circuit and Antenna Design for Communications Engineering*, Norwood, MA: Artech House, 2006.

[4] Warnick, K. F., and P. Russer, *Problem Solving in Electromagnetics, Microwave Circuit and Antenna Design for Communications Engineering*, Norwood, MA: Artech House, 2006.

5

ENERGY AND ELECTRIC POTENTIAL OF CHARGE SYSTEMS

In this chapter, the concept of work is introduced as applicable to the movement of charge in free space. Discussion of electric potential between two points is undertaken as is an introduction to the gradient. The relationship between field intensity, **E**, and the gradient, ∇ is presented. The energy in static electric fields is also discussed and quantified.

5.1 WORK DONE IN MOVING A POINT CHARGE

In an electric field **E**, a point charge, Q, experiences a force that is given by

$$\mathbf{F} = Q\mathbf{E} \quad \text{(N)} \tag{5.1}$$

This unbalanced force will result in an acceleration of the charged particle with its motion in the direction of the field if Q is positive [1] (see Figure 5.1).

To put the charge in equilibrium, an *applied force* is required; that is, an applied force that is equal in magnitude and opposite in direction to the force from the field [1]:

$$\mathrm{F}_a = -Q\mathbf{E} \quad \text{(N)} \tag{5.2}$$

Work is defined as a force acting over a distance [2]. Therefore, when the charge particle moves at constant speed through a differential distance, *dl*, the applied force does a differential amount of work, *dW*. The work may be positive or negative with respect to the applied force, $\mathbf{F}_a$, depending upon the direction of the vector displacement, $d\mathbf{l}$.

When $d\mathbf{l}$ and $\mathbf{F}_a$ are not in the same direction, the component of the force in the direction of $d\mathbf{l}$ must be used, expressed as:

$$dW = F_a dl \cos\theta = \mathbf{F_a} \cdot d\mathbf{l} \quad \text{(Joules)} \tag{5.3}$$

The differential work done by an external agent in an electric field is expressed as:

$$dW = -Q\mathbf{E} \cdot d\mathbf{l} \quad \text{(Joules)} \tag{5.4}$$

With this as the defining expression for work that is done in moving a charge particle in an electric field, a positive value means that work had to be done by an external agent in order to bring about the change in position. A negative result will mean that work was done by the field [3].

It is important to know the expressions for $d\mathbf{l}$ in the three coordinate systems [4]. These are:

$$d\mathbf{l} = dx\ \mathbf{x} + dy\ \mathbf{y} + dz\ \mathbf{z}; \quad \text{(Cartesian coordinates)} \tag{5.5}$$

$$d\mathbf{l} + dr\ \mathbf{r} + rd\phi\ \boldsymbol{\phi} + dz\ \mathbf{z}; \quad \text{(cylindrical coordinates)} \tag{5.6}$$

$$d\mathbf{l} = dr\ \mathbf{r} + r\ d\theta\ \boldsymbol{\theta} + r\sin\theta\ d\phi\ \boldsymbol{\phi}; \quad \text{(spherical coordinates)} \tag{5.7}$$

The work done in moving a point charge Q from point B to point A in a static electric field is the same for *any* path chosen. Equivalently, the work done in moving the charge around any closed loop is zero; that is,

$$\oint \mathbf{E} \cdot d\mathbf{l} = 0 \quad \text{(static fields)} \tag{5.8}$$

Such a vector field is called a *conservative* field [2].

5.2 ELECTRIC POTENTIAL BETWEEN TWO POINTS

The *potential* of point A with respect to point B is defined as the work done in moving a unit positive charge, Q_u, from point B to point A [1].

$$V_{AB} = \frac{W}{Q_u} = -\int \mathbf{E} \cdot d\mathbf{l} \quad \text{(J/C or V)} \tag{5.9}$$

The terms potential and voltage are interchangeable.

It should be observed also that the initial, or reference, point is the lower limit of the line integral. Then, too, the minus sign must not be omitted. This sign came into the expression by way of the force $\mathbf{F}_a = -Q\mathbf{E}$ that had to be applied to put the charge in equilibrium.

Because $\mathbf{E}$ is a conservative field,

$$V_{AB} = V_{AC} - V_{BC} \tag{5.10}$$

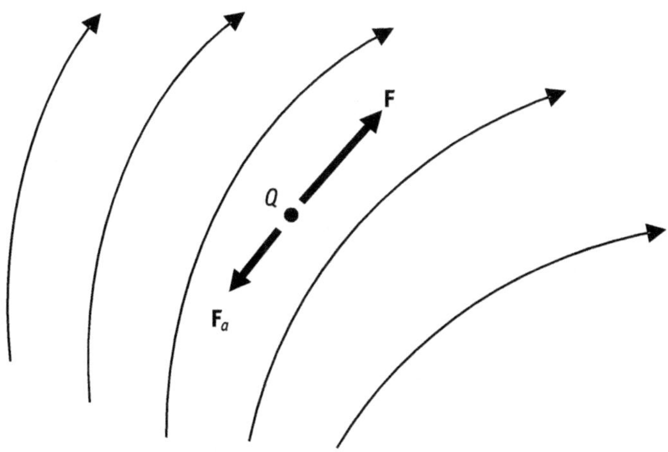

Figure 5.1 Force on a positive charge Q when moved from one point to another in an electric field.

V_{AB} may be considered now as the *potential difference* between points A and B. When V_{AB} is positive, work must be done to move the unit positive charge from point B to point A. Then, point A is said to be at a higher potential than point B. For example:

If a charge of +2C located at point B (2, 0, 0) on the graph in Figure 5.2, is moved to point A (0, 2, 0) and an energy of 24J is expended to do it, then

$$V_{AB} = \frac{Q}{W} = \frac{24\text{J}}{+2\text{C}} = +12\text{V} \tag{5.11}$$

where:

$Q = +2\text{C}$

$W = 24\text{J}$

It may be concluded then that point A is at a higher potential than B (i.e., 12V higher). Also, the potential V_{BA} must be −12V since V_{BA} differs from V_{AB} only by an interchange of the lower and upper limits in the defining integral, which simply changes the sign of the result.

5.3 POTENTIAL OF A POINT CHARGE

Since the electric field due to a point charge Q is completely in the radial direction, spherical coordinates are most conveniently used, and, moreover, only the radial (r-) component is needed [2]:

$$\begin{aligned} V_{AB} &= \int_{B}^{A} \mathbf{E} \cdot d\mathbf{l} = -\int_{r_B}^{r_A} E_r \, dr = -\left(\frac{Q}{4\pi\varepsilon_0}\right)\int_{r_B}^{r_A}\left(\frac{1}{r^2}\right)dr \\ &= \left(\frac{Q}{4\pi\varepsilon_0}\right)\left[\left(\frac{1}{r_A}\right)-\left(\frac{1}{r_B}\right)\right] \end{aligned} \tag{5.12}$$

For a positive charge Q, point A is at a higher potential than point B when r_A is smaller than r_B. The equipotential surfaces are concentric spherical shells.

If reference point B is now allowed to move out to infinity (∞), then:

$$V_{A\infty} = \left(\frac{Q}{4\pi\infty}\right)\left[\left(\frac{1}{r_A}\right) - \left(\frac{1}{\infty}\right)\right] \quad (5.13)$$

or

$$V = \left(\frac{Q}{4\pi\varepsilon_0 r}\right) \quad (5.14)$$

Considerable use will be made of this equation in the discussions that follow. The greatest danger to a problem-solution analyst is forgetting where the reference is while attempting to apply the equation to charge distributions which themselves extend to ∞. See Figure 5.2 for an illustration of work performed when a point charge is moved from one location to another.

5.4 POTENTIAL OF A CHARGE DISTRIBUTION

If the charge is distributed throughout some finite volume with a known charge density, ρ_V (unit: C/m^3), then the potential at some

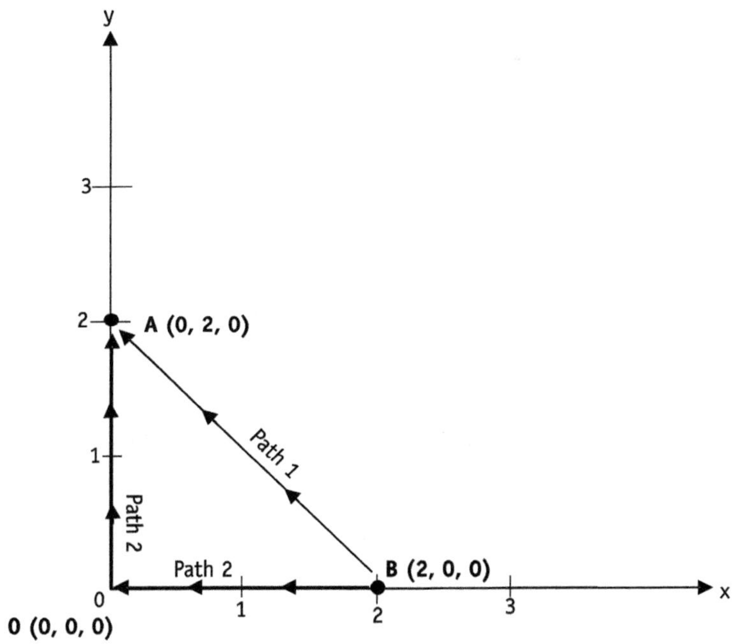

Figure 5.2 Illustration of work performed moving charge Q from one point to another in an electric field.

external point can be determined. To make this determination, a differential charge at an arbitrary point within the volume is identified (see Figure 5.3).

At point P:

$$dV = \frac{dQ}{4\pi\varepsilon_0 R} \tag{5.15}$$

Integration over the volume, *V*, gives the total potential (difference) at P.

Integral development:

$$dV = \frac{dQ}{4\pi\varepsilon_0 R} = \frac{(\rho_v\, dV)}{4\pi\varepsilon_0 R} \tag{5.16}$$

where $dQ = \rho_v\, dV$

$$\int_V dV = \int_V \frac{(\rho_v\, dV)}{4\pi\varepsilon_0 R} \tag{5.17}$$

to get:

$$V = \int \frac{(\rho_v\, dV)}{4\pi\varepsilon_0 R} = \int \left(\frac{\rho_v}{4\pi\varepsilon_0 R}\right) dV \tag{5.18}$$

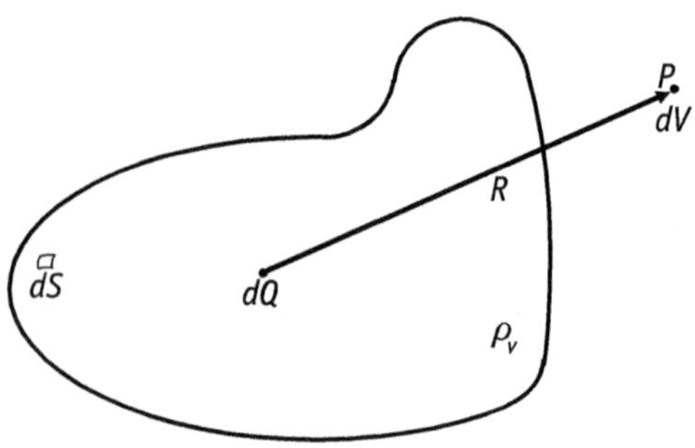

Figure 5.3 Arbitrary differential charge, *dQ*, in a volume of electric charge with charge density of ρ_v.

R must not be confused with r of the spherical coordinates system. It is not a vector but is the scalar distance from dQ to the fixed point, P. R always varies from place to place throughout the volume. As such, it cannot be removed from the integrand.

If charge is distributed over a surface or a line, the expression for V also holds, provided that integration is over the surface or the line with ρ_S or ρ_l used in place of ρ_V. It must be emphasized that all these expressions for the potential at an external point are based upon a *zero voltage (0V) reference at* ∞ [2].

5.5 GRADIENT

In this section, another vector analysis operation is introduced—*gradient.* Figure 5.4(a) shows two neighboring points of a region for which a scalar function is defined [3].

The vector separation of the two points is:

$$d\mathbf{r} = dx\mathbf{x} + dy\mathbf{y} + dz\mathbf{z} \tag{5.19}$$

From the calculus mathematics, the change in V from M to N is given by:

$$dV = \left(\frac{\partial V}{\partial x}\right)dx + \left(\frac{\partial V}{\partial y}\right)dy + \left(\frac{\partial V}{\partial x}\right)dz \tag{5.20}$$

Now the ∇ operator—introduced in Section 4.4—operating on V, gives:

$$\nabla V = \frac{\partial V}{\partial x}\mathbf{x} + \frac{\partial V}{\partial y}\mathbf{y} + \frac{\partial V}{\partial z}\mathbf{z} \tag{5.21}$$

It follows that:

$$dV = \nabla V \cdot d\mathbf{r} \tag{5.22}$$

The vector field ∇V—*also written as grad V*—is called the gradient of the scalar function V. It is seen that, for fixed $|d\mathbf{r}|$, the change in V in a given direction $d\mathbf{r}$ is proportional to the projection of ∇V in that direction. Thus, ∇V *lies in the direction of maximum increase of the function* V [2].

Another view of the gradient—see Figure 5.4(b)—is obtained by allowing the points M and N to lie on an equipotential surface, $V(x, y, z) = c_1$. Then $dV = 0$, which implies that ∇V is perpendicular to $d\mathbf{r}$. But $d\mathbf{r}$ is tangent to the equipotential surface. Indeed, for a suitable location of N, it represents *any* tangent through M. Therefore, ∇V must be along the surface normal at M. Since ∇V is in the direction of increasing V, it points from $V(x, y, z) = c_1$ to $V(x, y, z) = c_2$, where $c_2 > c_1$. Thus, the gradient of a potential function is a vector field that is everywhere normal to the equipotential surfaces.

The gradient in the cylindrical and spherical coordinate systems follows directly from that in the Cartesian system. It is noted

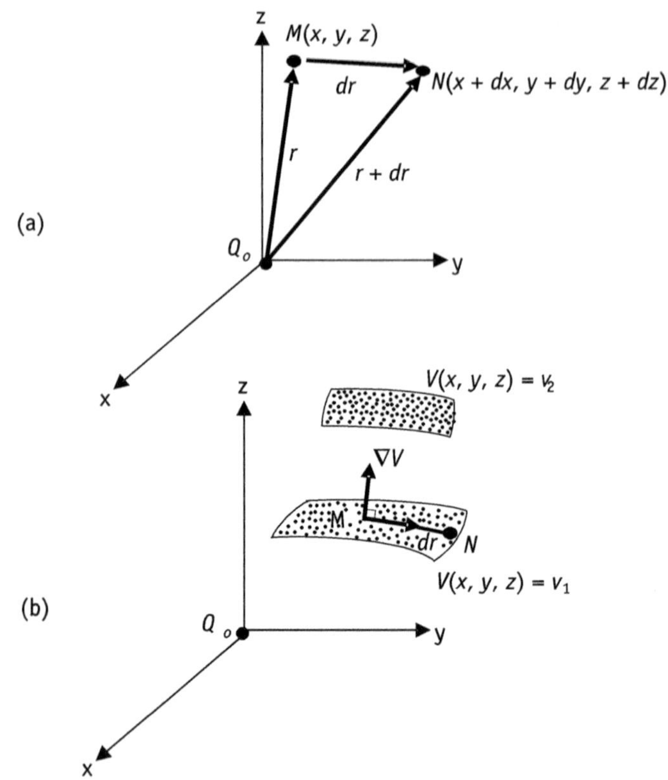

Figure 5.4 (a) Two neighboring points of a region in which a scalar function is defined, and (b) the same neighboring points on an equipotential surface.

that each term contains the partial derivative of V with respect to distance in the direction of that particular unit vector [4]:

$$\nabla V = \frac{\partial V}{\partial x}\mathbf{x} + \frac{\partial V}{\partial y}\mathbf{y} + \frac{\partial V}{\partial z}\mathbf{z} \quad \text{(Cartesian coordinates)} \tag{5.23}$$

$$\nabla V = \frac{\partial V}{\partial r}\mathbf{r} + \frac{\partial V}{r\,\partial \phi}\boldsymbol{\phi} + \frac{\partial V}{\partial z}\mathbf{z} \quad \text{(cylindrical coordinates)} \tag{5.24}$$

$$\nabla V = \frac{\partial V}{\partial r}\mathbf{r} + \frac{\partial V}{r\,\partial \theta}\boldsymbol{\theta} + \frac{\partial V}{[r\sin\theta\,\partial \phi]}\boldsymbol{\phi} \quad \text{(spherical coordinates)} \tag{5.25}$$

While ∇V is written for grad V in any coordinate system, it must be remembered that the ∇ operator is defined only in Cartesian coordinates [4].

5.6 RELATIONSHIP BETWEEN E AND ∇

From the integral expression for the potential of A with respect to B, the differential of V may be written

$$dV = -\mathbf{E} \cdot d\mathbf{l} \tag{5.26}$$

On the other hand,

$$dV = \nabla V \cdot d\mathbf{r} \tag{5.27}$$

Since $d\mathbf{l} = d\mathbf{r}$ is an arbitrary small displacement, it follows that:

$$\mathbf{E} = -\nabla V \tag{5.28}$$

The electric field intensity **E** may be obtained when the potential function V is known by simply taking the negative of the gradient of V. The gradient was found to be a vector normal to the equipotential surfaces, directed to a positive change in V. With the negative sign, the **E** field is found to be directed from higher to lower levels of the potential V [2].

5.7 ENERGY IN STATIC ELECTRIC FIELDS

Consider the work required to assemble, charge by charge, a distribution of $n = 3$ point charges. The region is assumed initially to be charge-free and with $\mathbf{E} = 0$ throughout. Referring to Figure 5.5, the work required to place the first charge, Q_1, into position 1 is zero.

Then, when Q_2 is moved toward the region, work equal to the product of this charge and the potential due to Q_1 is required. The total work to position the three charges is:

$$\begin{aligned} W_E &= W_1 + W_2 + W_3 \\ &= 0 + \left(Q_2 V_{2,1}\right) + \left(Q_3 V_{3,1} + Q_3 V_{3,2}\right) = W_{E1} \end{aligned} \tag{5.29}$$

The potential $V_{2,1}$ reads, "The potential at point 2 due to charge Q_1 at position 1." (Note: This unique notation will not appear again in the remainder of this book.) The work, W_E, is the energy stored in the electric field of the charge distribution.

If the three charges were brought into place in reverse order, the total work would be:

$$\begin{aligned} W_E &= W_3 + W_2 + W_3 \\ &= 0 + \left(Q_2 V_{2,3}\right) + \left(Q_1 V_{1,3} + Q_1 V_{1,2}\right) = W_{E2} \end{aligned} \tag{5.30}$$

When the two expressions, (5.29) and (5.30), are added, the result is twice the stored energy [2]:

$$\begin{aligned} 2W_E &= W_{E1} + W_{E2} \\ &= \left(W_1 + W_2 + W_3\right) + \left(W_3 + W_2 + W_1\right) \\ &= \left[0 + \left(Q_2 V_{2,1}\right) + \left(Q_3 V_{3,1} + Q_3 V_{3,2}\right)\right] \\ &\quad + \left[0 + \left(Q_2 V_{2,3}\right) + \left(Q_1 V_{1,3} + Q_1 V_{1,2}\right)\right] \\ &= Q_1\left(V_{2,1} + V_{1,3}\right) + Q_2\left(V_{2,1} + V_{2,3}\right) + Q_3\left(V_{3,1} + V_{3,2}\right) \end{aligned} \tag{5.31}$$

where

$W_{E1} = W_1 + W_2 + W_3;$
$W_{E2} = W_3 + W_2 + W_1;$
$W_1 + W_2 + W_3 = 0 + (Q_2V_{2,1}) + (Q_3V_{3,1} + Q_3V_{3,2});$
$W_3 + W_2 + W_1 = 0 + (Q_2V_{2,3}) + (Q_1V_{3,1} + Q_1V_{1,2}).$

The term $Q_1(V_{1,2} + V_{1,3})$ was the work done against the fields of Q_2 and Q_3, the only other charges in the region. Hence, $V_{1,2} + V_{1,3} = V_1$, the potential at position 1. Then,

$$\begin{aligned} 2W_E &= Q_1\left(V_{1,2} + V_{1,3}\right) + Q_2\left(V_{2,1} + V_{2,3}\right) + Q_3\left(V_{3,1} + V_{3,2}\right) \\ &= Q_1V_1 + Q_2V_2 + Q_3V_3 \end{aligned} \tag{5.32}$$

where

$V_1 = V_{1,2} + V_{1,3};$
$V_2 = V_{2,1} + V_{2,3};$
$V_3 = V_{3,1} + V_{3,2};$

and

$$W_E = \left(\frac{1}{2}\right) \sum_{m=3}^{n} Q_m V_m \tag{5.33}$$

for a region containing n point charges. For a region with a charge density, ρ_v (C/m^3), the summation becomes an integration [3]:

$$W_E = \left(\frac{1}{2}\right) \int \rho_v V \, dv \tag{5.34}$$

Other useful forms of the expression for stored energy are:

$$W_E = \left(\frac{1}{2}\right) \int \mathbf{D} \cdot \mathbf{E} \, dv \tag{5.35}$$

$$W_E = \left(\frac{1}{2}\right) \int \left(\varepsilon E^2\right) dv \tag{5.36}$$

and

$$W_E = \left(\frac{1}{2}\right)\int\left(\frac{D^2}{\varepsilon}\right)dv \tag{5.37}$$

In an electric circuit, the energy stored in the field of a capacitor is given by:

$$W_E = \left(\frac{1}{2}\right)QV = \left(\frac{1}{2}\right)CV^2 \tag{5.38}$$

where:

- C is the capacitance (in Farads, F);
- V is the voltage difference between the two conductors making up the capacitor (in volts, V);
- Q is the magnitude of the total charge on one of the conductors (in coulombs, C).

5.8 PROBLEMS AND SOLUTIONS: ENERGY AND ELECTRIC POTENTIAL OF CHARGE SYSTEMS

Problem 5.1

Find the work done moving a charge of +2C from (2, 0, 0)m to (0, 2, 0)m along the straight line path joining the two points if the electric field is:

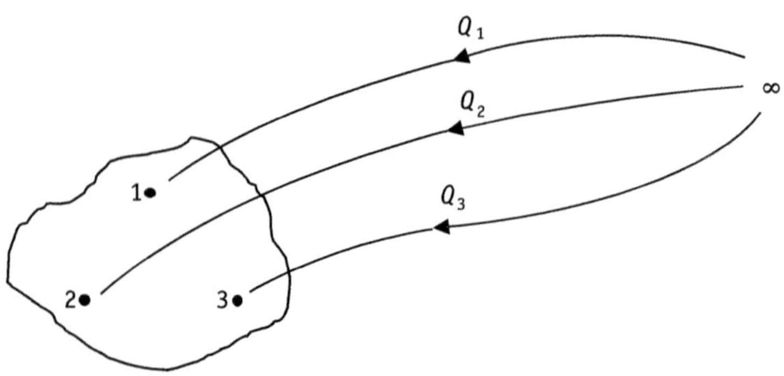

Figure 5.5 Work performed by moving charges Q_1, Q_2, and Q_3 from ∞ to points 1, 2, and 3.

$$\mathbf{E} = 2x\mathbf{x} - 4y\mathbf{y} \text{ (V/m)}$$

Problem 5.2

Find the work done in the field of the previous example problem when the +2C charge is moved from (2, 0, 0)m to (0, 0, 0)m along the x-axis and then from (0, 0, 0) to (0, 2, 0) along the y-axis.

Problem 5.3

Find the potential of A(1, ϕ, z) with respect to B(3, ϕ, z) in cylindrical coordinates where the electric field due to a line charge on the z-axis as given by $\mathbf{E} = 50\ (1/r)\mathbf{r}$ (V/m).

Problem 5.4

Find the work performed by an external agent applying a force to move a point charge of $Q = -20\ \mu$C from the origin to (4, 0, 0) in the field:

$$\mathbf{E} = (\tfrac{1}{2}x + 2y)\mathbf{x} + 2xy\mathbf{y} \text{ (V/m)}$$

Problem 5.5

In the field of Problem 5.1, move the charge from (4, 0, 0)m to (4, 2, 0)m and determine the work performed.

Problem 5.6

In the E-field of Problem 5.1, find the work performed by moving the charge from the origin to (4, 2, 0) along the straight line connecting the points.

Problem 5.7

Find the work done in moving a point charge of $Q = 5\ \mu$C from the origin to the spherical coordinate point (2m, $\pi/4$, $\pi/2$) in the field

$$\mathbf{E} = 5\ e^{-(1/4)r}\ \mathbf{r} + (10/r^2 \sin\theta)\ \boldsymbol{\varphi} \text{ (V/m).}$$

Problem 5.8

Given the field $\mathbf{E} = (k/r)\mathbf{r}$ in cylindrical coordinates, show that the work needed to move a point charge Q from any radial distance, r, to a point at twice that radial distance is independent of that distance.

Problem 5.9

For a line charge, ρ_l, = ½ × 10^9 (C/m) on the z-axis, find V_{AB} where point A is located at (2m, $\pi/2$, 0) and point B is at (4m, π, 5m).

Problem 5.10

In the field of Problem 5.9, find V_{BC} where r_B = 4m and r_c = 10m. Then, find V_{AC} (r_A = 2m) and compare with the sum of V_{AB} and V_{BC}.

Problem 5.11

Find the potential of *pt* (2, π, $\pi/2$) with respect to (4, 0, π) given the field:

$$\mathbf{E} = (-16/r^2)\ \mathbf{r}\ (\text{V/m}).$$

Problem 5.12

A line charge ρ_l = 400 pC/m lies along the x-axis and the surface of zero potential passes through the point P_A(0, 5, 12)m in Cartesian coordinates. Find the potential at P_B(2, 3, −4)m.

Problem 5.13

Find the potential at r_A = 5m with respect to r_B = 15m due to a point charge Q = 500 pC at the origin and zero reference at infinity.

Problem 5.14

A total charge of (40/3)nC is uniformly distributed around a circular ring of radius 2m.

(a) Find the potential at point on the axis 5m from the plane of the ring;
(b) Compare with the result where the full charge is at the origin in the form of a point charge.

Problem 5.15

A total charge of (40/3)nC is uniformly across a circular disk radius, r. What is the potential when r is:

(a) 1m;
(b) 2m;

(c) 4m;

(d) 8m.

Problem 5.16

Five equal point charges of $Q = 20\ \mu$C each are located at $X = 2, 3, 4, 5$, and 6m.

Find the potential at the origin.

References

[1] Cheng, D. K., *Fundamentals of Engineering Electromagnetics*, Reading, MA: Addison-Wesley Publishing Company, Inc., 1993.

[2] Shadowitz, A., *The Electromagnetic Field*, New York: Dover Publications, Inc., 1975.

[3] Russer, P., *Electromagnetics, Microwave Circuit and Antenna Design for Communications Engineering*, Norwood, MA: Artech House, 2006.

[4] Boyce, W. E., *Calculus*, New York: John Wiley & Sons, 1988.

6

CURRENT, CURRENT DENSITY, AND CONDUCTORS

Many electrostatic concepts have been introduced in the previous five chapters. In this chapter, the subject of electric current is tackled. Along with it are Ohm's law and what occurs when charges are in motion. Conduction current density and convection current density are discussed along with the concepts of conductivity, resistance, and the continuity of current. Finally, dielectric/conductor boundary conditions in physical materials are discussed.

6.1 INTRODUCTION

Electric current is the rate of transport of electric charge past a specified point or across a specified surface. The symbol, I, is generally used for constant currents. The symbol, i, is for time-variable currents. The unit of current is the *ampere (A)* (1A = 1 C/s). In the SI unit, the ampere is the basic unit while the coulomb is the derived unit [1].

Ohm's law relates current to voltage and resistance. For simple dc circuits, $I = V/R$. However, when charges are suspended in a liquid or a gas, or where both positive and negative charge carriers are present with different characteristics, the simple form of Ohm's law is insufficient. Consequently, the current density J (A/m^2) receives more attention in electromagnetics than does current I [2].

6.2 CHARGES IN MOTION

Consider the force on a positively charged particle in a vacuum and in an electric field as shown in Figure 6.1(a). This force,

$$\mathbf{F} = +Q\mathbf{E} \quad \text{(Newtons)} \tag{6.1}$$

is unopposed and results in constant acceleration. Thus, the charge moves in the direction of **E** with a velocity **U** that increases as long as the particle is in the **E** field.

When the charge is in a liquid or gas, as shown in Figure 6.1(b), it collides repeatedly with particles in the medium, resulting in random changes in direction. But, for constant **E** and a homogeneous medium, the random velocity components cancel out, leaving a constant average velocity, known as the *drift velocity*, **U**, along the direction of **E** [1].

Conduction in metals takes place by movement of the electrons in the outermost shells of the atoms that make up the crystalline structure. According to *electron-gas theory*, these electrons reach an

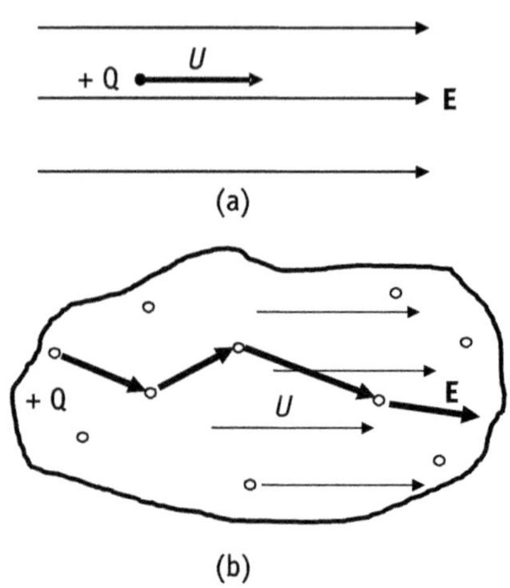

Figure 6.1 Force on positively-charged particle in: (a) vacuum, and (b) liquid or gas.

average drift velocity in much the same way as a charged particle moving through a liquid or gas. The drift velocity is directly proportional to the electric field intensity,

$$\mathbf{U} = \mu\mathbf{E} \quad (\text{m}^2/\text{V-s}) \tag{6.2}$$

The *mobility*, μ, has the units m^2/V-s. Each cubic meter of a conductor contains on the order of 10^{28} atoms. Good conductors have one or two electrons from each atom free to move upon application of the field [2].

The mobility, μ, varies in temperature and the crystalline structure of the solid. The particles in the solid have a vibratory motion. The vibratory motion increases with temperature, making it more difficult for the charges to move. Thus, at higher temperatures mobility, μ, is reduced, resulting in a smaller drift velocity (or current) for a given **E**. In circuit analysis, this phenomenon is accounted for by stating a *resistivity* for each material and specifying an increase in the resistivity with increasing temperature [1].

6.3 CONVECTION CURRENT DENSITY, *J*

A set of charged particles giving rise to a charge density, ρ_v, in a volume, *v*, is shown to have a velocity **U** to the right as can be seen in Figure 6.2 [3].

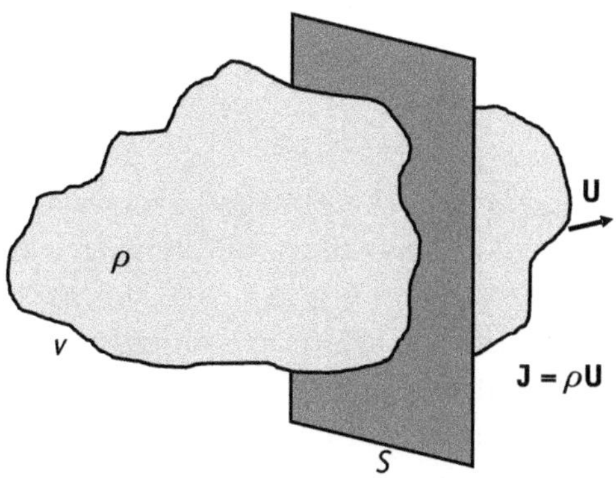

Figure 6.2 Illustration of a convection current.

The particles are assumed to maintain their relative positions within the volume. As this charge configuration passes a surface, *S*, it constitutes a *convection current* with density:

$$\mathbf{J} = \rho\mathbf{U} \quad (\mathrm{A/m^2}) \tag{6.3}$$

Of course, if the cross section of *v* varies or if the density ρ is not constant throughout *v*, then **J** will not be constant with time. Furthermore, **J** will be zero when the last portion of the volume crosses *S*. Nevertheless, the concept of a current density caused by a cloud of charged particles in motion is at times useful in the study of electromagnetic field theory [1].

6.4 CONDUCTION CURRENT DENSITY, *J*

Of more interest is the *conduction current* that occurs in the presence of an electric field within a conductor of fixed cross-section. The current density is again given by:

$$\mathbf{J} = \rho\mathbf{U} \quad (\mathrm{A/m^2}) \tag{6.4}$$

which, in view of the relation $\mathbf{U} = \mu\mathbf{E}$, can be written as

$$\mathbf{J} = \sigma\mathbf{E} \quad (\mathrm{A/m^2}) \tag{6.5}$$

where

$\sigma = \rho\mu$;

σ is the conductivity of the material in *siemens per meter (S/m)* [4].

In metallic conductors the charge carriers are electrons. These electrons (i.e., negative charge carriers), drift in a direction opposite to that of the electric field (see Figure 6.3). Hence, for electrons (negative charge carriers), both ρ and μ are negative. This in turn results in a positive conductivity, σ, just as in the case of positive charge carriers. Hence, it follows that vectors **J** and **E** have the same direction regardless of the sign of the charge carriers [2].

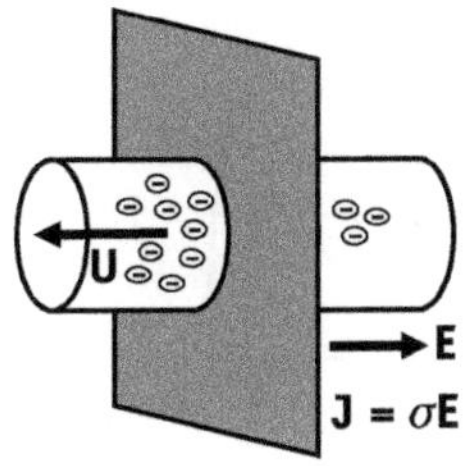

Figure 6.3 Illustration of a conduction current.

It is conventional to treat electrons as moving to the left, and positive charge carriers as moving to the right, always reporting ρ and μ as positive [3].

The relation, $\mathbf{J} = \sigma\mathbf{E}$, is often referred to as the *point form of Ohm's law*. The factor, σ, takes into account the density of the electrons being free to move (ρ) and the relative ease with which they move through the crystalline structure (μ). As might be expected, σ is also a function of temperature [1].

6.5 CONDUCTIVITY, Σ

In a liquid or gas, there are generally both positive and negative ions present, some being singly charged and others being doubly charged with possibly different masses. A conductivity expression would include all such factors. However, it is assumed that all the negative ions are alike and likewise the positive ions, then the conductivity contains two terms as shown in Figure 6.4(a) [3].

In a metallic conductor, only the valence electrons are free to move. In Figure 6.4(b) they are shown in motion going to the left. The conductivity then contains only one term, the product of the charge density of the electrons free to move, ρ_e, and their mobility, μ_e [2]. A somewhat more complex conduction occurs in semiconductor materials such as germanium and silicon. In the crystal structure, each atom has four covalent bonds with adjacent atoms.

However, at room temperature and upon influx of energy from some external source, such as light, electrons can move out of the position called for by covalent bonding to create an *electron-hole pair* available for electrical conduction. Such materials are call *intrinsic* semiconductors. Electron-hole pairs have a short lifetime,

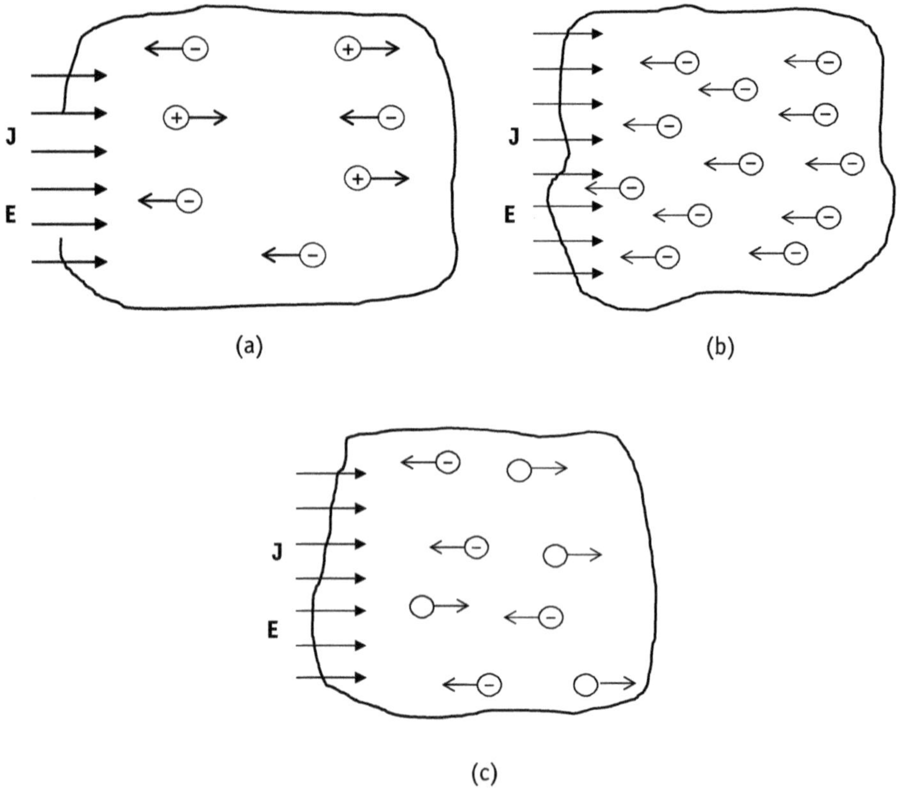

Figure 6.4 E-field and charge carrier movement in various matter types: (a) liquid or gas, (b) metallic conductor, and (c) semiconductor.

disappearing by way of *recombination*. However, other pairs are being formed constantly. Thus, at all times, some are available for conduction.

As shown in Figure 6.4(c), the conductivity, σ, consists of two terms, one for the electrons and another for the holes [4]. In practice, impurities, in the form of valence-three or valence-five elements, are added to create *p-type* and *n-type* semiconductor materials, respectively. The intrinsic behavior just described continues but is far overshadowed by the presence of extra electrons in n-type materials or holes in p-type materials. Then, in the conductivity σ, one of the densities, ρ_e or ρ_h, will far exceed the other [2].

6.6 CURRENT, *I*

The total current, I (in A), crossing a surface S is given by:

$$I = \mathbf{J} \cdot d\mathbf{S} \quad \text{(A)} \tag{6.6}$$

See Figure 6.5.

A normal vector must be chosen for the differential surface $d\mathbf{S}$. The positive result for I indicates a current crossing S in the direction of this normal vector. Of course, $\mathbf{J}$ need not be uniform over S, and S need not be a planar surface [1].

6.7 RESISTANCE, *R*

If a conductor of uniform cross-sectional area, A, and length, l, as shown in Figure 6.6, has a voltage difference, V, between its ends, then:

$$E = \frac{V}{l} \qquad \text{(V/m)} \tag{6.7}$$

and

$$J = \sigma\left(\frac{V}{l}\right) = \sigma E \quad \text{(A/m}^2\text{)} \tag{6.8}$$

where $E = V/l$.

Assuming that the current is uniformly distributed over the area A. The total current, I, is then:

$$I = JA = \frac{(\sigma AV)}{l} \quad \text{(Amperes, A)} \tag{6.9}$$

Since Ohm's law states that $V = IR$, the resistance is:

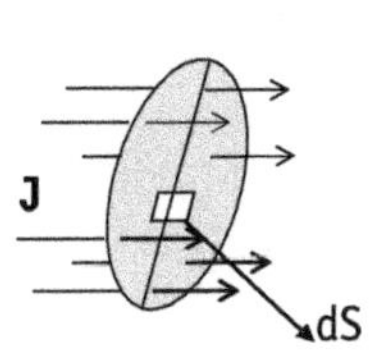

Figure 6.5 Total current crossing a surface.

$$R = \frac{l}{(\sigma A)} \quad (\Omega) \tag{6.10}$$

(Note that 1 $S^{-1} = 1\Omega$. The siemens was formerly known as the *mho*.)

This expression for resistance is generally applied in all conductors where the cross-section remains constant over the length, l. However, if the current density is greater along the surface area of the conductor than in the center, then the expression is not valid. For such nonuniform current distribution, the resistance is given by:

$$R = \frac{V}{\int(\mathbf{J} \cdot d\mathbf{S})} = \frac{V}{\int(\sigma\mathbf{E} \cdot d\mathbf{S})} \quad (\Omega) \tag{6.11}$$

If **E** is known rather than the voltage difference between the two faces, the resistance is given by:

$$R = \frac{\int(\mathbf{E} \cdot d\mathbf{l})}{\int(\sigma\mathbf{E} \cdot d\mathbf{S})} \quad (\Omega) \tag{6.12}$$

The numerator gives the voltage drop across the sample, while the denominator gives the total current, I [2].

6.8 CURRENT SHEET DENSITY, *K*

At times current is confined to the surface of a conductor, such as inside the walls of a waveguide. For such a current sheet, it is helpful

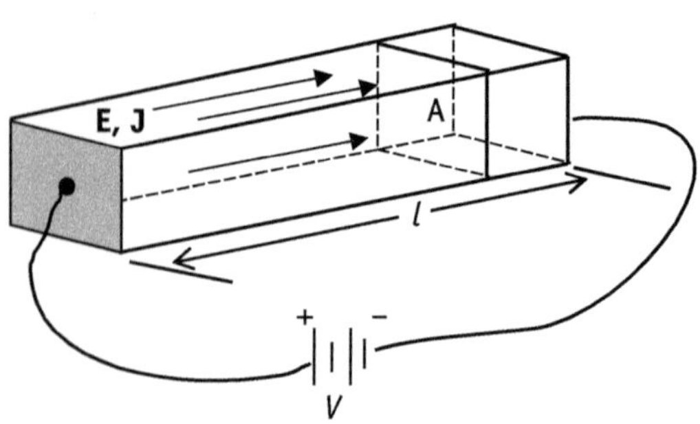

Figure 6.6 Conductor of uniform cross-sectional area with voltage potential applied between ends.

to define the density vector, **K** (in A/m), which gives the rate of charge transport per unit transverse length (note: some reference books use the notation $\mathbf{J}_S$). Figure 6.7 shows a total current of I in the form of a cylindrical sheet of radius, r, flowing in the positive z direction.

In this case,

$$\mathbf{K} = \left(\frac{I}{2\pi r}\right)\mathbf{a}_z = \left(\frac{I}{2\pi r}\right)\mathbf{z} \quad \text{(A/m)} \tag{6.13}$$

at each point of the sheet. For other sheets, **K** might vary from point to point. In general, the current flowing through a curve C within a current sheet is obtained by integrating the *normal* component of **K** along the curve (see Figure 6.8).

Thus,

$$I = \int_C K_n \, dl \quad \text{(A)} \tag{6.14}$$

6.9 CONTINUITY OF CURRENT

Current, I, crossing a general surface S has been examined where **J** at the surface was known. Now, if the surface is *closed* in order for net current to come out [5], there must be a decrease of positive charge within:

$$\oint \mathbf{J} \cdot d\mathbf{S} = I = -\frac{dQ}{dt} = -\frac{\partial}{\partial t}\int \rho \, dv \quad \text{(A)} \tag{6.15}$$

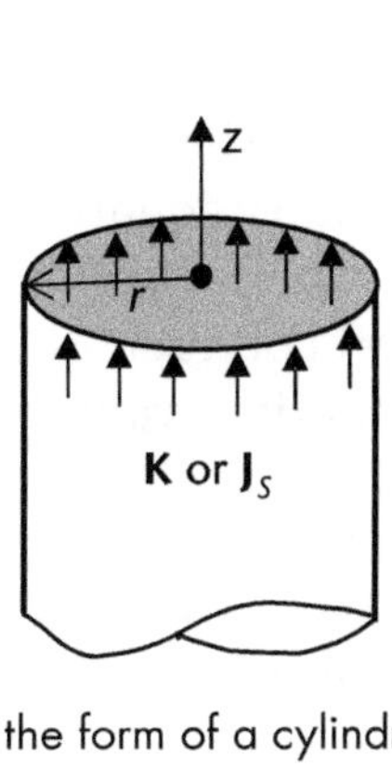

Figure 6.7 Total current, I, in the form of a cylindrical sheet of radius, r.

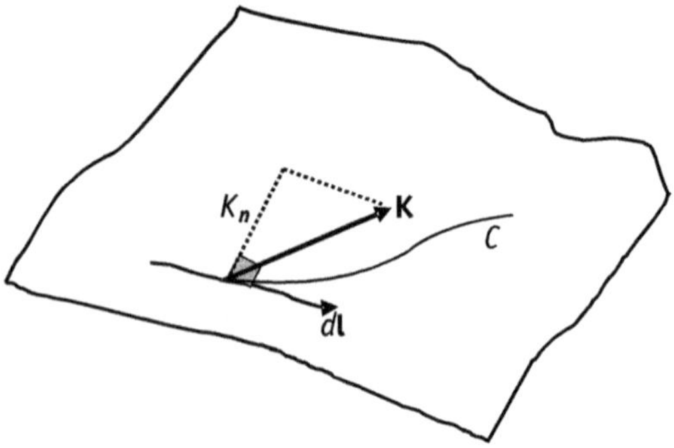

Figure 6.8 Current flowing through a curve C within a current sheet.

where the unit normal in $d\mathbf{S}$ is the outward-directed normal. Dividing by Δv,

$$\frac{\left(\oint \mathbf{J} \cdot d\mathbf{S}\right)}{\Delta v} = -\frac{\partial}{\partial t}\left[\frac{\left(\int \rho \, dv\right)}{\Delta v}\right] \tag{6.16}$$

As $\Delta v \to 0$, the left side by definition approaches $\nabla \cdot \mathbf{J}$, the divergence of the current density, while the right side approaches $-\partial\rho/\partial t$. Thus,

$$\nabla \cdot \mathbf{J} = -\frac{\partial \rho}{\partial t} \tag{6.17}$$

Equation (6.17) is the *continuity of current* equation. In it ρ stands for the *net charge* density, not just the density of mobile charge. As will be shown below, $\partial\rho/\partial t$ can be nonzero within a conductor only transiently. Then, the continuity equation, $\nabla \cdot \mathbf{J} = 0$, becomes the field equivalent of Kirchoff's current law, which states that the net current leaving a junction of several conductors is zero.

In the process of conduction, valence electrons are free to move upon the application of an electric field. So, to the extent that these electrons are in motion, static conditions no longer exist. However, these electrons should not be confused with *net charge*, for each conduction electron is balanced by a proton in the nucleus such that there is zero net charge in every Δv of the material. Suppose, however, that through a temporary imbalance a region within a solid conductor has a *net* charge density ρ_0 at time $t = 0$. Then, since

$$\mathbf{J} = \sigma\mathbf{E} = \left(\frac{\sigma}{\varepsilon}\right)\mathbf{D} \quad (\mathrm{A/m^2}) \tag{6.18}$$

$$\nabla \cdot \left(\frac{\sigma}{\varepsilon}\right)\mathbf{D} = -\frac{\partial\rho}{\partial t} \tag{6.19}$$

The divergence operation consists of partial derivatives with respect to the spatial coordinates. If σ and ε are constants—as they would be in a homogeneous sample—then they may be removed from the partial derivatives.

$$\left(\frac{\sigma}{\varepsilon}\right)(\nabla \cdot \mathbf{D}) = -\frac{\partial\rho}{\partial t} \tag{6.20}$$

$$\left(\frac{\sigma}{\varepsilon}\right)\rho = -\frac{\partial\rho}{\partial t} \quad \text{or} \quad \left(\frac{\sigma}{\varepsilon}\right)\rho + \frac{\partial\rho}{\partial t} = 0 \tag{6.21}$$

The solution to the equation is:

$$\rho = \rho_0 e^{-t/\tau} = \rho_0 e^{-t/(\varepsilon/\sigma)} = \rho_0 e^{-(\sigma/\varepsilon)t} \tag{6.22}$$

where $\tau = \varepsilon/\sigma$ and $\varepsilon = \varepsilon_r\varepsilon_0$.

This becomes useful when knowing the relative dielectric constant of the material of interest. Also when determining charge dissipation (sometimes called charge relaxation) in liquid Jet A commercial airplane fuel that has already been filled in airplane fuel tanks. However, such fuel also has conductive (liquid) additives to increase the Jet A fuel's conductivity, σ.

It is seen that ρ, and with it

$$\frac{\partial\rho}{\partial t} = -\left(\frac{\sigma}{\varepsilon}\right)\rho \tag{6.23}$$

decays exponentially with a time constant, $\tau = \varepsilon/\sigma$, also known as the *relaxation time* for the particular material. For silver (Ag), having $\sigma = 6.17 \times 10^7$ S/m and $\varepsilon = \varepsilon_0$, the relaxation time is 1.44×10^{-19} s. Thus, if a charge density of ρ_0 could somehow be set up in the interior of a block of silver, the charges would separate due to repelling coulomb forces. After 1.44×10^{-19} s, the remaining density would be 36.8% of ρ_0. After five time-constants (or 7.20×10^{-19} s), only 0.67% of ρ_0 remains. Thus, for static fields, it may be said that the *net charge within a conductor is zero.* If any net charge is present, it must reside on the outer surface [1].

6.10 CONDUCTOR: DIELECTRIC BOUNDARY CONDITIONS

Under static conditions, all net charge will be on the outer surface of a conductor, and both **E** and **D** are therefore zero within the conductor [4]. Because the electric field is a conservative field, the closed-line integral of $\mathbf{E} \cdot d\mathbf{l}$ is zero for any path. A rectangular path with corners 1, 2, 3, 4 is shown in Figure 6.9.

Analysis:

$$\underset{1}{\int_1^2 \mathbf{E} \cdot d\mathbf{l}} + \underset{2}{\int_2^3 \mathbf{E} \cdot d\mathbf{l}} + \underset{3}{\int_3^4 \mathbf{E} \cdot d\mathbf{l}} + \underset{4}{\int_4^4 \mathbf{E} \cdot d\mathbf{l}} = 0 \quad \text{(V/m)} \tag{6.24}$$

If path length 4-to-1 and path length 2-to-3 are now allowed to approach zero while keeping the interface between them, then the second and fourth integrals in the above equation are also equal to zero. The path length from 3-to-4 is within the conductor where **E** must be zero. This leaves:

$$\int_1^2 \mathbf{E} \cdot d\mathbf{l} = \int_1^2 Et\, dl = 0$$

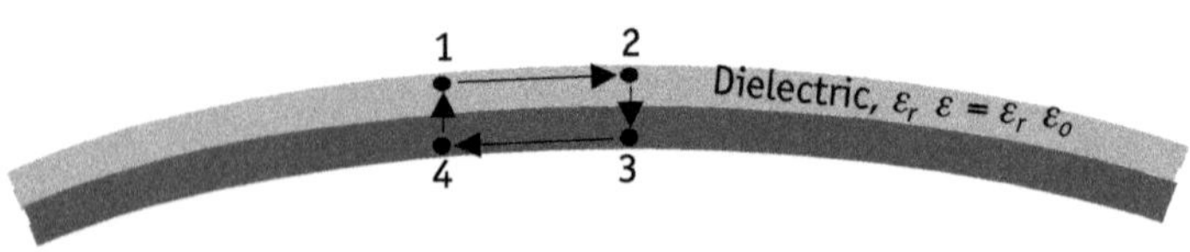

Figure 6.9 Illustration of rectangular path, adjacent dielectric, and conductor analysis: boundary conditions.

where E_t = the tangential component of **E** at the dielectric surface.

Since the interval 1-to-2 can be chosen arbitrarily,

$$E_t = D_t = 0 \tag{6.25}$$

at each point of the surface.

To discover the conditions on the normal components, a small, closed, right-circular cylinder is placed across the interface as shown in Figure 6.10.

Gauss' law applied to this surface gives:

$$\oint \mathbf{D} \cdot d\mathbf{S} = Q_{\text{enc}} \tag{6.26}$$

$$\int_{(\text{top})} \mathbf{D} \cdot d\mathbf{S} + \int_{(\text{bottom})} \mathbf{D} \cdot d\mathbf{S} + \int_{(\text{side})} \mathbf{D} \cdot d\mathbf{S} = \int_{A} \rho_S \, dS \tag{6.27}$$

The third integral is zero since, as just determined, $D_l = 0$ on either side of the interface. The second integral is also zero, since the bottom of the cylinder is within the conductor, where **D** and **E** are zero. Then,

$$\int_{(\text{top})} \mathbf{D} \cdot d\mathbf{S} + \int_{(\text{top})} D_n \cdot dS + \int_{A} \rho_S \, dS \tag{6.28}$$

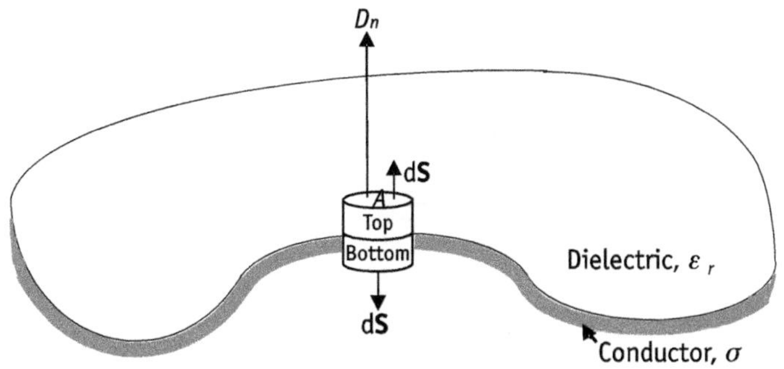

Figure 6.10 Small, closed, right-circular cylinder placed across a dielectric-conductor interface.

which can hold only if:

$$D_n = \rho_S \tag{6.29}$$

and

$$E_n = \frac{\rho_S}{\varepsilon} \tag{6.30}$$

In short, under static conditions the field just outside a conductor is zero (both tangential and normal components) unless there exists a surface charge distribution [2]. However, a surface charge does not imply a net charge in the conductor. To illustrate, consider a positive charge at the origin of spherical coordinates. Now if this point charge is enclosed by an *uncharged* conducting spherical shell of finite thickness as shown in Figure 6.11, then the field is still given by:

$$\mathbf{E} = \left(\frac{+Q}{4\pi\varepsilon r^2}\right)\mathbf{r} \quad (\text{V/m}) \tag{6.31}$$

except within the conductor itself, where **E** must be zero. The coulomb forces caused by $+Q$ attract the conduction electrons to the inner surface where they create a ρ_{s1} of negative sign. Then, the deficiency of electrons on the outer surface constitutes a positive surface charge

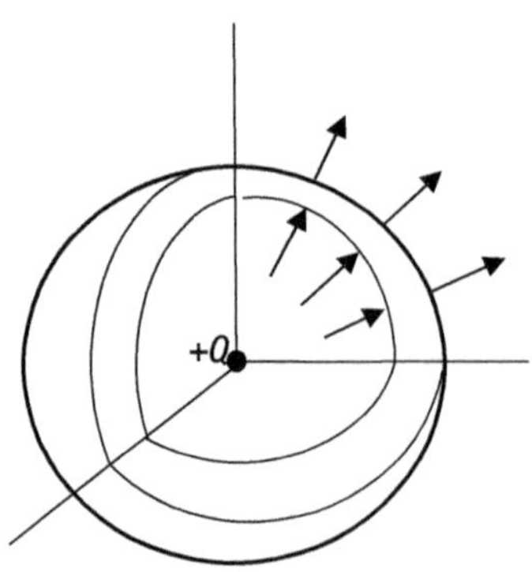

Figure 6.11 Point charge enclosed by uncharged conducting shell of finite thickness.

density, ρ_{s2}. The electric flux lines, Ψ, leaving the point charge $+Q$, terminate at the electrons on the inner surface of the conductor [4] as shown in Figure 6.12.

Then, electric flux lines Ψ originate once again on the positive charges on the outer surface of the conductor. It should be noted that the flux does not pass through the conductor, and the *net* charge on the conductor remains zero [4].

6.11 PROBLEMS AND SOLUTIONS: CURRENT, CURRENT DENSITY, AND CONDUCTORS

Problem 6.1

Find I in the circular wire (see Figure 6.13) of radius 2 mm if the current density, $\mathbf{J}$, is:

$$\mathbf{J} = 15\,(1 - e^{-1000r})\,\mathbf{z} \quad (\text{A/m}^2)$$

Problem 6.2

Find the resistance, R, between the inner and outer curved surfaces of the block shown in Figure 6.14. The material of the block is silver (Ag, $\sigma = 6.17 \times 10^7$ S/m).

Problem 6.3

An AWG #12 Cu conductor has an 80.8-mm diameter. A 50-ft length carries a current of 20A.

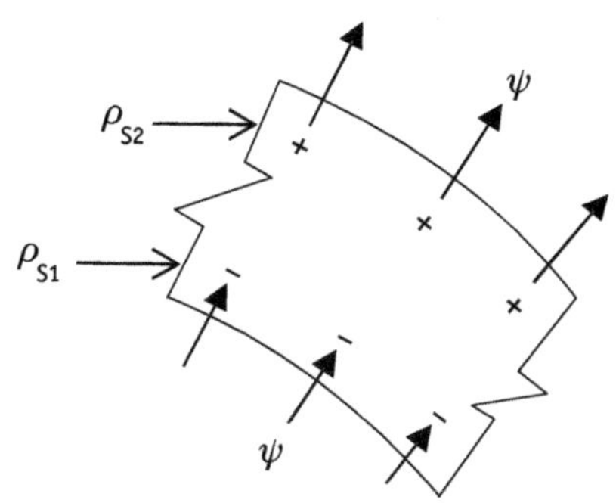

Figure 6.12 Electric flux lines, Ψ, leaving point charge $+Q$, terminating at electrons on the conductor inner surface to, once again, originate on positive charges on the conductor outside surface.

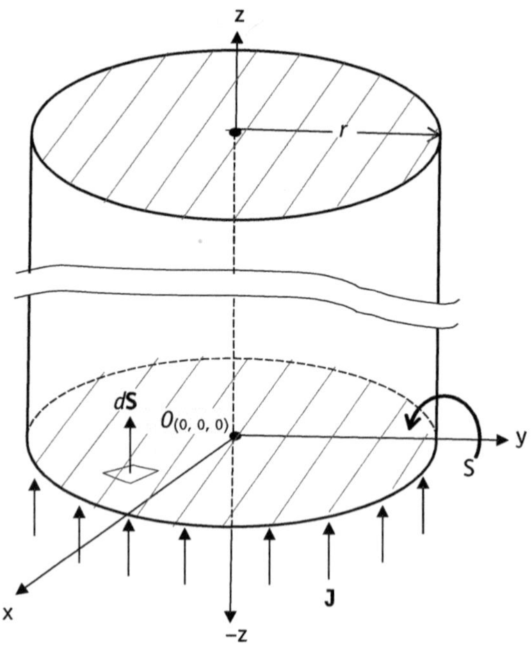

Figure 6.13 Illustration of vector analysis for Problem 6.1.

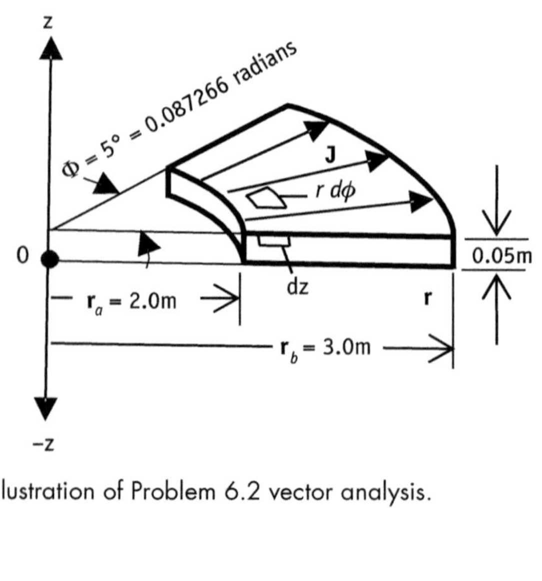

Figure 6.14 Illustration of Problem 6.2 vector analysis.

(a) What is E, the electric field intensity?

(b) What is U, the drift velocity?

(c) What is the voltage drop, V?

(d) What is the resistance, R, of the 50-ft conductor?

Problem 6.4

What current density, J, and electric field intensity, E_o, correspond to a drift velocity of 5.3×10^{-4} m/s?

Problem 6.5

A long Cu conductor has a circular cross section of diameter 3.0 mm and carries a 10A current. Each second, what percentage of the conduction electrons must leave to be replaced by others in a 100-mm length?

Problem 6.6

What current would result if all the conduction electrons in a 1-cm cube of Al passed a specified point in 2.0s? Assume one conduction electronic per atom.

Problem 6.7

What is the density of free electrons in a metal for an electron mobility, μ_M, of 0.0046 m^2/Vs and a conductivity, σ, of 29.1 MS/m?

Problem 6.8

Determine the conductivity of intrinsic Ge at room temperature.

Problem 6.9

Find the conductivity of n-type Ge at room temperature. Assume one donor atom in each batch of 10^8 atoms. The density of Ge is 5.32×10^3 kg/m^3, and the atomic weight is 72.6 kg/kmole.

Problem 6.10

A conductor of uniform cross-section and 150m length has a voltage drop of 1.3V and a current density of 4.65×10^5 A/m^2.

What is the conductivity of the material in the conductor?

Problem 6.11

A table of resistivities gives 10.4Ω CM/ft for annealed Cu.

What is the corresponding conductivity in S/m?

Problem 6.12

An AWG #20 Al wire has a resistance of $16.7\ \Omega/1000$ ft. What conductivity for Al does this imply?

Problem 6.13

Find the current, I, crossing the portion of the $y = 0$ plane defined by:

$$-0.1\text{m} \le x \le 0.1\text{m and } -0.002\text{m} \le z \le 0.002\text{m if:}$$

$$\mathbf{J} = 10^2|x|\mathbf{y} \quad (\text{A/m}^2)$$

Problem 6.14

Find the current, I, crossing the portion of the $x = 0$ plane defined by:

$$-\pi/4 \le y \le \pi/4$$

and

$$-0.01\text{m} \le z \le 0.01\text{m}$$

if:

$$\mathbf{J} = 10^2 \cos(2y)\mathbf{x} \quad (\text{A/m}^2).$$

Problem 6.15

Find the current, I, crossing the spherical shell with $r = 0.02$m given:

$$\mathbf{J} = 10^3 \sin\theta\, \mathbf{r} \quad (\text{A/m}^2).$$

Problem 6.16

Show that the resistance of any conductor of constant cross-sectional area, A, and length, l, is given by:

$$R = l/\sigma A,$$

assuming uniform current distribution.

Problem 6.17

A current sheet of a 4m-width lies in the $z = 0$ plane and contains a total I of 10A in a direction from the origin to $P_1(1, 3, 0)$m.

Problem 6.18

For the configuration contained in the previous problem, find the current sheet in a 30° sector of the plane.

References

[1] Shadowitz, A., *The Electromagnetic Field*, New York: Dover Publications, Inc., 1975.

[2] Russer, P., *Electromagnetics, Microwave Circuit and Antenna Design for Communications Engineering*, Norwood, MA: Artech House, 2006.

[3] Warnick, K. F., and P. Russer, *Problem Solving in Electromagnetics, Microwave Circuit and Antenna Design for Communications Engineering*, Norwood, MA: Artech House, 2006.

[4] Cheng, D. K., *Fundamentals of Engineering Electromagnetics*, Reading, MA: Addison-Wesley Publishing Company, Inc., 1993.

[5] Boyce, W. E., *Calculus*, New York: John Wiley & Sons, 1988.

7

CAPACITANCE AND DIELECTRIC MATERIALS

7.1 POLARIZATION *P* AND RELATIVE PERMITTIVITY

Dielectric materials become *polarized* in an electric field with the result that their electric flux density, D, being greater than would be under free-space conditions and under the same field intensity.

A simplified—but satisfactory—theory of polarization treats an atom of a dielectric material with two superimposed positive and negative charge regions as shown in Figure 7.1(a).

Upon applying an **E** field, the positive charge region moves in the direction of the applied field with the negative charge region moving in the direction opposite the field as illustrated in Figure 7.1(b). As a result, this displacement can be represented by an *electric dipole moment*,

$$\mathbf{p} = Q\mathbf{d} \tag{7.1}$$

as illustrated in Figure 7.1(c).

For most materials, when the applied field is removed, the charge regions will return to their original superimposed positions. Just as in the case of a spring obeying Hooke's law, the work done in the atom distortion is recovered when the system is permitted to go

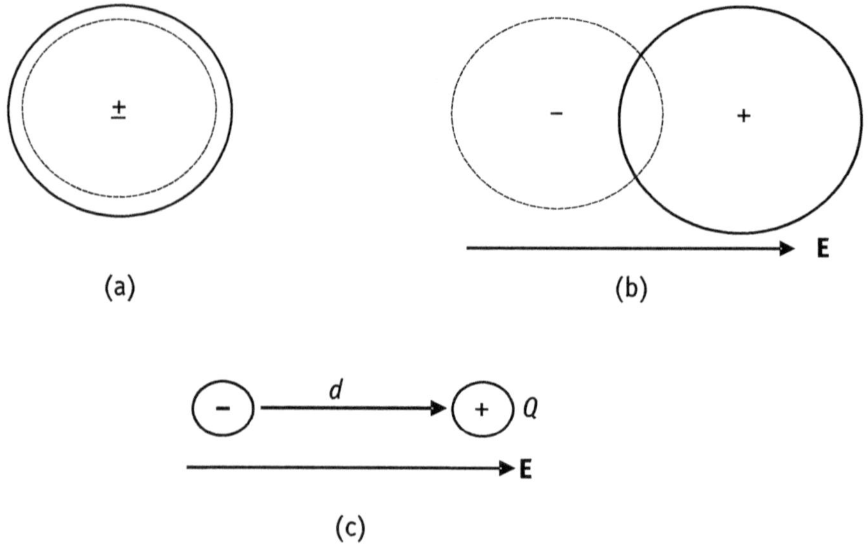

Figure 7.1 Illustration of a simplified but satisfactory theory of polarization for a dielectric: (a) atom, two superimposed, (b) positive charge region moving, and (c) resulting electric dipole positive and negative charge in the direction of the applied E-field, moment, $\mathbf{p} = Q\mathbf{d}$ regions negative-charge region moving oppositely.

back to its original material state. Energy storage taking place in this distortion is done in the same manner as with the spring.

A region Δv of a polarized dielectric will contain N-dipole moments of polarization, **p**.

Polarization **P** is defined as the dipole moment per unit volume:

$$P = \lim_{\Delta v \to 0}\left(\frac{N\mathbf{p}}{\Delta v}\right) \quad (\mathrm{C/m^2}) \tag{7.2}$$

This definition suggests, of course, a smooth and continuous distribution of electric dipole moments throughout the volume, and this is not the case. In the macroscopic view, the polarization **P** can account for the increase in the electric flux density as shown in the following equation:

$$\mathbf{D} = \varepsilon_0\mathbf{E} + \mathbf{P} \tag{7.3}$$

In this equation, **E** and **P** are permitted to have different directions, as they do in certain crystalline dielectrics. In an isotropic, linear material, **E** and **P** are parallel at each point. This occurrence is expressed by:

$$\mathbf{P} = \chi_e \varepsilon_0 \mathbf{E} \qquad \text{(isotropic material)} \tag{7.4}$$

Electric susceptibility, χ_e, is a dimensionless constant. Thus, we also have:

$$\mathbf{D} = \varepsilon_0 \left(1 + \chi_e\right)\mathbf{E} = \varepsilon_0 \varepsilon_r \mathbf{E} \qquad \text{(isotropic material)} \tag{7.5}$$

where $\varepsilon_r = 1 + \chi_e$ is also a pure number. Since, $\mathbf{D} = \varepsilon\mathbf{E}$,

$$\varepsilon_r = \frac{\varepsilon}{\varepsilon_0} \tag{7.6}$$

ε_r is called the *relative permittivity* [1] (compare Section 2.1).

7.2 FIXED VOLTAGE *D* AND *E*

A parallel-plate capacitor with free space between the plates and a constantly applied voltage, V, as shown in Figure 7.2, has a constant electric field intensity, **E**, between the plates.

Neglecting edge-fringing, three equations are helpful:

$$\mathbf{E} = \frac{V}{d}\mathbf{a}_n \tag{7.7}$$

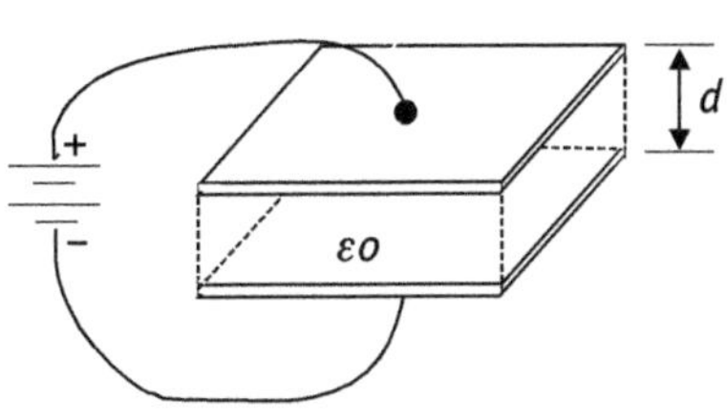

Figure 7.2 Capacitor with dielectric material between the plates and an applied voltage source.

$$\mathbf{D} = \varepsilon_0 \mathbf{E} = \varepsilon_0 \left(\frac{V}{d} \right) \mathbf{a}_n \tag{7.8}$$

$$D_n = \rho_s = \frac{Q}{A} \tag{7.9}$$

So, when a dielectric with relative permittivity, ε_r, fills the space between the plates,

$$\mathbf{D} = \varepsilon_0 \mathbf{E} + \mathbf{P} = \varepsilon_0 \mathbf{E} + \varepsilon_0 \chi_e \mathbf{E} \tag{7.10}$$

the equations are:

$$\mathbf{E} = \frac{V}{d} \mathbf{a}_n \qquad \text{(as in free space)} \tag{7.11}$$

$$\mathbf{D} = \varepsilon_0 \varepsilon_r \mathbf{E} \tag{7.12}$$

Since $D_n = \rho_s = Q/A$, the charge and charge density increase by the factor, ε_r, over their free-space values. This charge increase is supplied by the source voltage, V [2].

7.3 FIXED CHARGE *D* AND *E*

The parallel-plate capacitor in Figure 7.3 has a charge $+Q$ on the upper plate and $-Q$ on the lower plate.

The charges could have been the result of a connected voltage source, V, which was subsequently removed. With free space between the plates and fringing neglected,

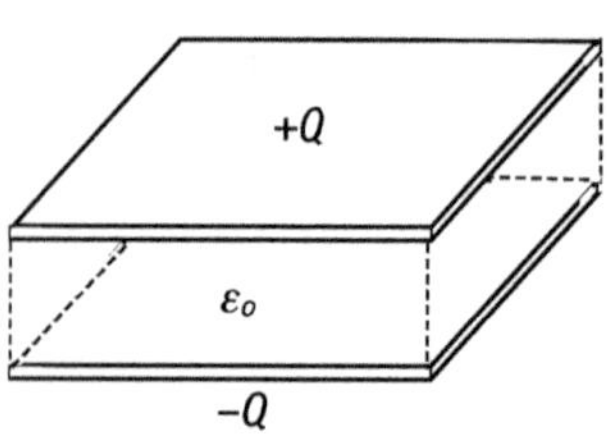

Figure 7.3 Capacitor with dielectric material between the plates and positive charge on the top plate and negative charge at the bottom plate.

$$D_n = \rho_s = \frac{Q}{A} \tag{7.13}$$

$$\mathbf{E} = \frac{\mathbf{D}}{\varepsilon_0} = \left(\frac{\rho_s}{\varepsilon_0}\right)\mathbf{a}_n \tag{7.14}$$

In this arrangement, there is no way for the charge to increase or decrease since, ideally (i.e., ideal dielectric material), there is no conducting path to the plates. When a dielectric material is assumed to fill the space between the plates, the equations are:

$$D_n = \rho_s = \frac{Q}{A} \tag{7.15}$$

$$\mathbf{E} = \frac{\mathbf{D}}{\varepsilon_0 \varepsilon_r} \tag{7.16}$$

With a constant Q and ρ_s, **D** must be the same as under free-space conditions whereas the magnitude of **E** decreases by the factor, $1/e_r$. The decrease in $\mathbf{E}\ \varepsilon_0$ is made up for by polarization, **P**, in the relation, $\mathbf{D} = \varepsilon_0\mathbf{E} + \mathbf{P}$. More generally, a homogeneous dielectric of relative permittivity, ε_r, the coulomb force between the charges is reduced to $1/e_r$ of its free-space value [1]:

$$\mathbf{F} = \left(\frac{Q_1 Q_2}{4\pi\varepsilon d^2}\right)\mathbf{a} = \left(\frac{1}{\varepsilon_r}\right)\left(\frac{Q_1 Q_2}{4\pi\varepsilon_0 d^2}\right)\mathbf{a} \tag{7.17}$$

7.4 BOUNDARY CONDITIONS AT THE INTERFACE OF TWO DIELECTRICS

If the conductor in Figures 6.9 and 6.10 is replaced by a second, and different dielectric, then the same argument, as was made in Section 6.10, can be made. In doing so, the following two boundary conditions are established:

(a) The tangential components of **E** are continuous across a dielectric interface. In symbols, this is:

$$E_{t1} = E_{t2} \tag{7.18}$$

and

$$\frac{D_{t1}}{\varepsilon_{r1}} = \frac{D_{t2}}{\varepsilon_{r2}} \tag{7.19}$$

(b) The normal component of **D** has a discontinuity of magnitude $|\rho_s|$ across a dielectric interface. If the normal unit vector is chosen to point into dielectric 2—from dielectric 1—then the following conditions can be written as:

$$D_{n1} - D_{n2} = -\rho_s \tag{7.20}$$

and

$$\varepsilon_{r1}E_{n1} - \varepsilon_{r2}E_{n2} = -\frac{\rho_s}{\varepsilon_0} \tag{7.21}$$

Generally speaking, the interface will have no free charges in it ($\rho_s = 0$) such that:

$$D_{t1} = D_{t2} \tag{7.22}$$

and

$$\varepsilon_{r1}E_{n1} = \varepsilon_{r2}E_{n2} \tag{7.23}$$

7.5 CAPACITANCE

Any two conducting bodies, separated by a dielectric material or free space (no dielectric material), have a capacitance that exists between them. A voltage difference applied between the two conductive bodies results in a $+Q$ charge on one conductor and $-Q$ on the other. The ratio of the absolute value of the charge to the absolute value of the voltage potential difference is defined as the capacitance of the system:

$$C = \frac{Q}{V} \text{ (Farads, F)}, \tag{7.24}$$

where 1 Farad (F) = 1 C/V.

The capacitance depends only on the geometry of the system and the properties of the dielectric(s) involved. In Figure 7.4, $+Q$ charge is placed on the upper conductor (#1), and $-Q$ charge is placed on the lower conductor (#2). The result is that a flux (E-) field is created. Both **D** and **E** fields are established together. Doubling both $+Q$ and $-Q$ simply doubles the magnitude of **D** and **E**, and doubles also the voltage potential difference between the conductive plates. Hence, the ratio Q/V would remain fixed [2].

7.6 MULTIPLE-DIELECTRIC CAPACITORS

When two dielectrics are present with the interface parallel to **E** and **D** as in Figure 7.5(a), the capacitance can be found by treating the arrangement as two capacitors in parallel (see Figure 7.5(a, b)).

Of course, the result may be extended to any number of side-by-side dielectrics. Hence, the equivalent capacitance is the sum of the two individual capacitances.

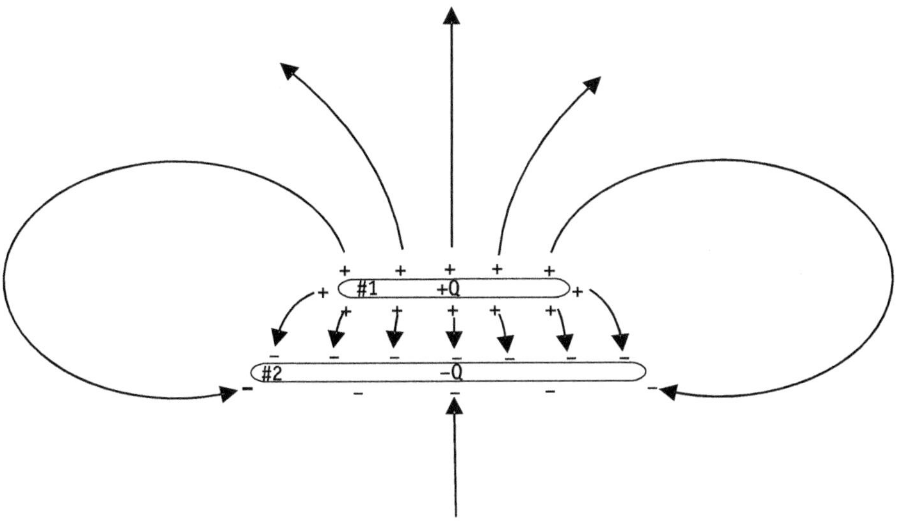

Figure 7.4 Capacitor: two plates of unequal width separated by a distance, showing paths and the direction of the E-fields when voltage is applied.

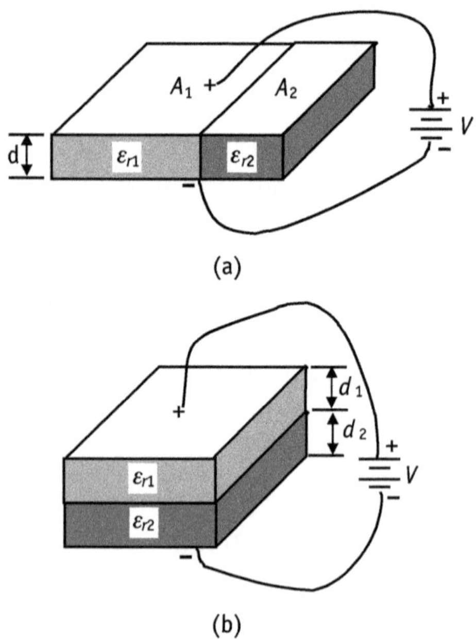

Figure 7.5 Capacitor example, two adjacent dielectric materials arranged in two different ways: (a) with the interface parallel and (b) with the interface normal.

When the dielectric interface is normal to **D** and E, as in Figure 7.5(b), the capacitance can be found by treating the arrangement as two capacitors in series, hence:

$$\frac{1}{C_{\text{equiv}}} = \frac{1}{C_1} + \frac{1}{C_2} \tag{7.25}$$

The result can be extended to any number of stacked dielectrics:

The reciprocal of the equivalent capacitance is the sum of the reciprocals of the individual capacitances [2].

7.7 ENERGY STORED IN A CAPACITOR

The energy stored in a capacitor is given by the expression:

$$W_E = \frac{1}{2}\int \mathbf{D} \cdot \mathbf{E}\, dv \tag{7.26}$$

Where the integration may be taken over is the space between the conductors (with the fringing neglected). If this space is occupied by a dielectric of relative permittivity, ε_r, then

$$\mathbf{D} = \varepsilon_0 \mathbf{E} + \mathbf{P} = \varepsilon_0 \varepsilon_r \mathbf{E} \tag{7.27}$$

and so

$$W_E = \frac{1}{2}\int \left(\varepsilon_0 E^2 + \mathbf{P} \cdot \mathbf{E}\right)\ dv = \frac{1}{2}\int \varepsilon_0 \varepsilon_r E^2\, dv \tag{7.28}$$

The two above expressions show how the presence of a dielectric results in an increase in stored energy over the free-space value ($\mathbf{P} = 0$, $\varepsilon_r = 1$) either through the term $\mathbf{P} \cdot \mathbf{E}$ or through the factor $\varepsilon_r > 1$ [4]. Given in terms of the capacitance,

$$W_E = \frac{1}{2} C V^2 \tag{7.29}$$

and the effect of the dielectric is reflected in C which is directly proportional to ε_r [1].

7.8 PROBLEMS AND SOLUTIONS: CAPACITANCE AND DIELECTRIC MATERIALS

Problem 7.1

Find the capacitance of a coaxial capacitor of length, h, where the outer surface of the inner conductor has a radius, a, and the inner surface of the outer conductor has a radius, b. See Figure 7.6.

Problem 7.2

Find the capacitance of the parallel plates (see Figure 7.7). Neglect fringing.

Problem 7.3

Find the polarization $\mathbf{P}$ in a dielectric material with $e_r = 2.8$ if:

$$\mathbf{D} = 3.0 \times 10^{-7}\ \mathbf{a} \quad \text{C/m}^2$$

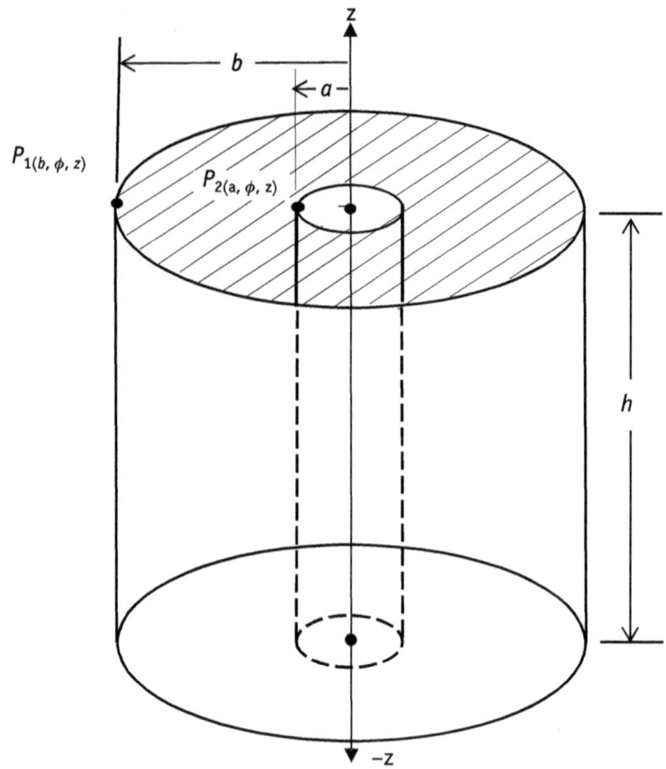

Figure 7.6 Illustration of vector analysis for Problem 7.1.

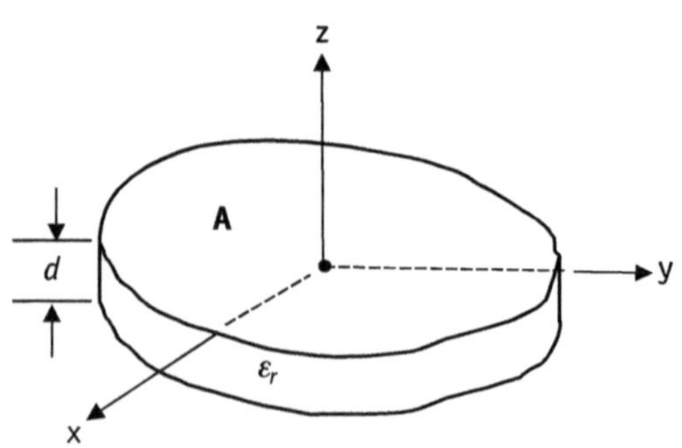

Figure 7.7 Parallel plates with dielectric material in between them.

Problem 7.4

Determine the value of **E** in a material with an electric susceptibility of 3.5 and:

$$\mathbf{P} = 2.30 \times 10^{-7}\ \mathbf{a} \quad \mathrm{C/m^2}.$$

Problem 7.5

Two point charges in a dielectric medium ($e_r = 5.2$) interact with a force of 8.6×10^{-3} N. What force could be expected if the charges were in free space?

Problem 7.6

Region 1, shown in Figure 7.8, is defined by $x < 0$ and is in free space. Region 2 in the figure, defined by $x > 0$, is a dielectric material with an $\varepsilon_r = 2.4$.

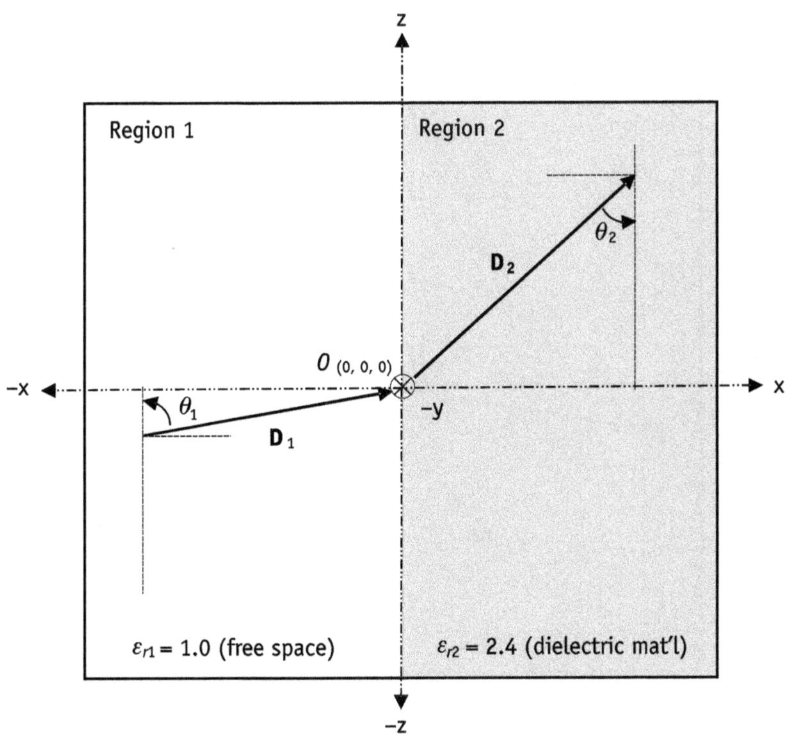

Figure 7.8 Free space/dielectric material interface, region 1 and region 2.

Given $\mathbf{D}_1 = 3\mathbf{x} - 4\mathbf{y} + 6\mathbf{z}$, find $\mathbf{E}_2$, find angles θ_1 and θ_2.

Problem 7.7

In the free-space region (#1) where $x < 0$,

$$\mathbf{E}_1 = 3\mathbf{x} + 5\mathbf{y} - 3\mathbf{z} \quad \text{V/m}.$$

In the region (#2) where $x > 0$ is a dielectric material of $\varepsilon_r = 3.6$. Find θ_2 that the E-field makes in the dielectric material with the $x = 0$ plane (see Figure 7.9).

Problem 7.8

A dielectric-free space interface (i.e., a plane) is characterized by the equation:

$$3x + 2y + z = 12\text{m}.$$

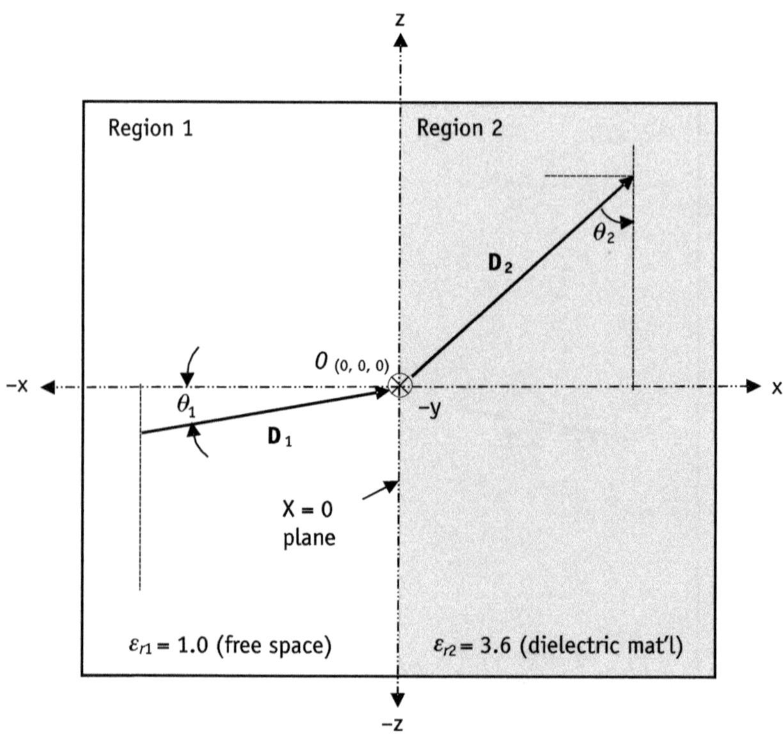

Figure 7.9 Free-space/dielectric material interface for Problem 7.7.

The origin side of the interface has $\varepsilon_r = 3.0$ and an E-field of:

$$\mathbf{E}_1 = 2\mathbf{x} + 5\mathbf{z} \quad \text{V/m}$$

Find E_2. See Figure 7.10.

Problem 7.9

Figure 7.11 shows a planar dielectric slab with free space on either side. Show that $\mathbf{E}_3 = \mathbf{E}_1$, assuming a constant field $\mathbf{E}_2$ with the slab.

Problem 7.10

Show that the capacitor of Figure 7.12 has capacitance:

$$C_{eq} = \varepsilon_0\varepsilon_{r1}\,(A_1/d) + \varepsilon_0\varepsilon_{r2}\,(A_2/d) = C_1 + C_2$$

Problem 7.11

Show that the capacitor of Figure 7.13 has reciprocal capacitance,

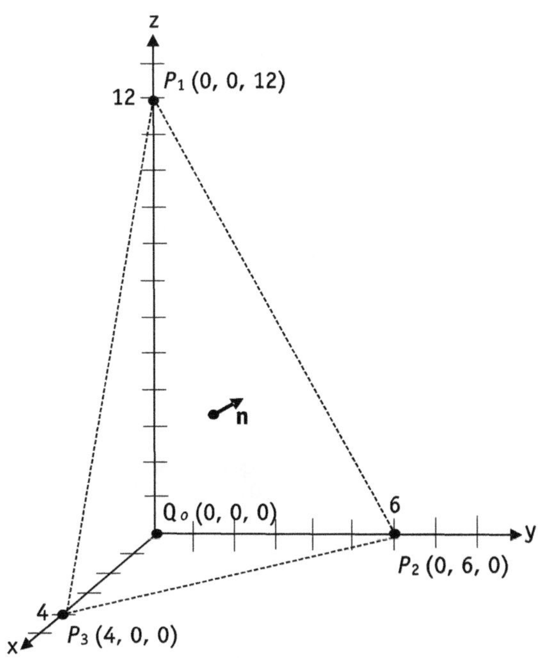

Figure 7.10 Dielectric-free space interface for Problem 7.8.

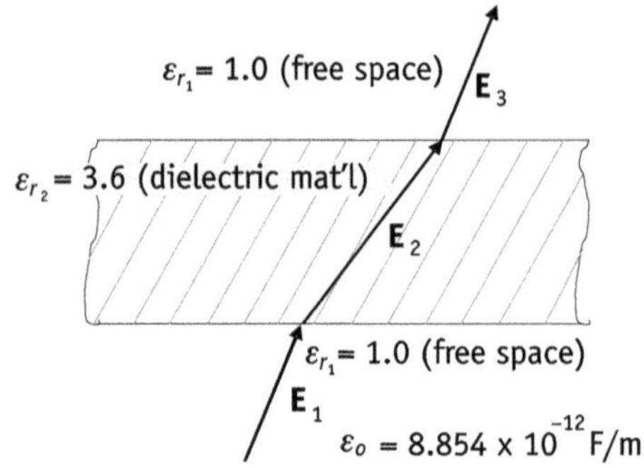

Figure 7.11 Free-space/dielectric material interface for Problem 7.9.

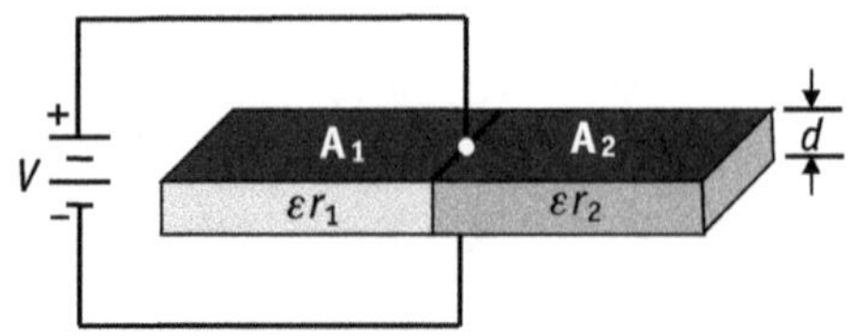

Figure 7.12 Free-space/dielectric material interface for Problem 7.10.

$$1/C_{eq} = 1/[\varepsilon_0\varepsilon_{r1}(A_1/d)] + 1/[\varepsilon_0\, \varepsilon_{r2}\, (A_2/d)] = 1/C_1 + 1/C_2$$

Problem 7.12

Find the capacitance of a coaxial capacitor of length, L, where the inner conductor has a radius, a, and the outer conductor has a radius, b (see Figure 7.14).

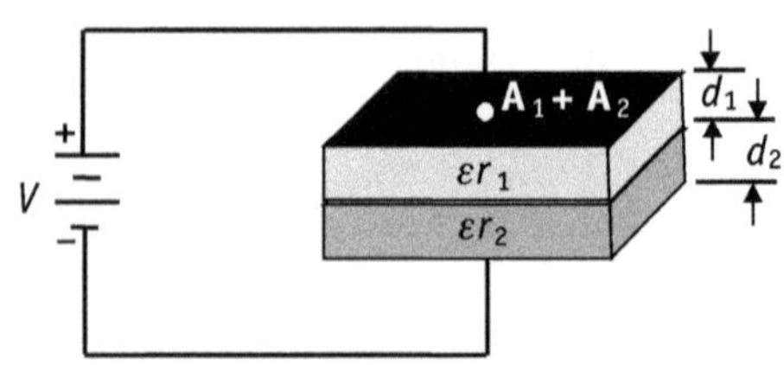

Figure 7.13 Free-space/dielectric material interface for Problem 7.11.

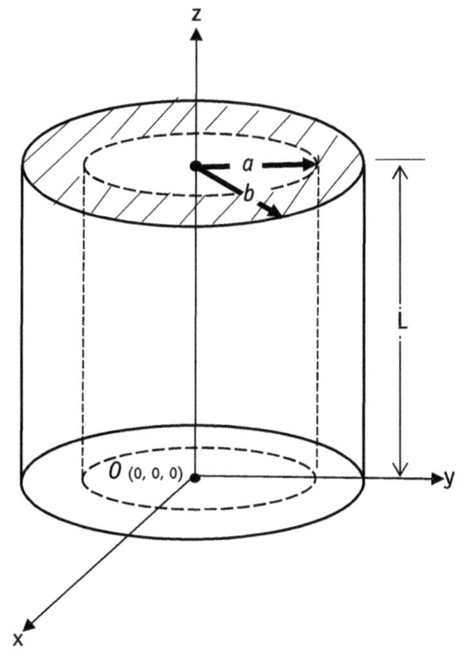

Figure 7.14 Free-space/dielectric material interface between cylindrical conductors.

Problem 7.13

In the capacitor of Figure 7.15, the region between the plates is filled with a dielectric of $\varepsilon_r = 4.5$.

Find the capacitance (see Figure 7.15).

Problem 7.14

Find the capacitance of an isolated spherical shell of radius, a (see Figure 7.16).

Problem 7.15

Find the capacitance between the two spherical shells of radius, a, separated by a distance, $d >> a$ (see Figure 7.17).

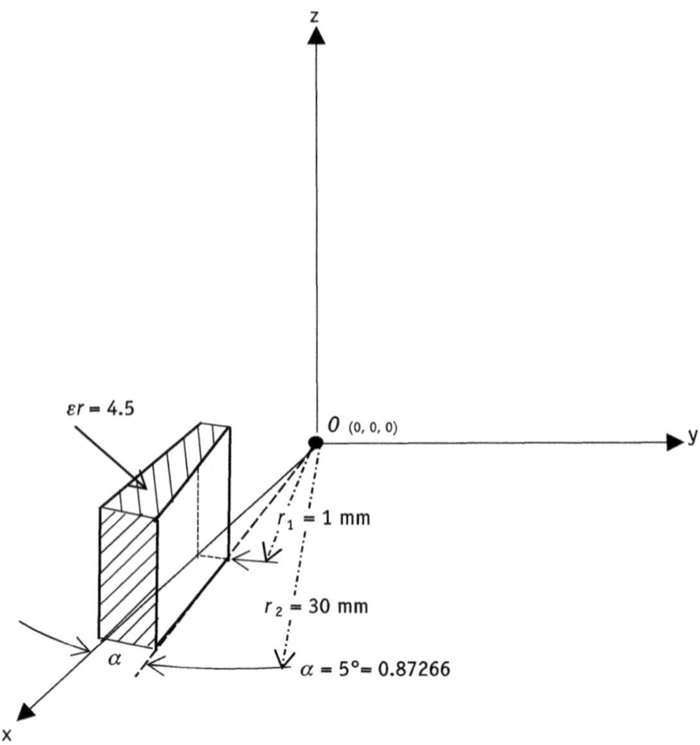

Figure 7.15 Free-space/dielectric material interface for Problem 7.13.

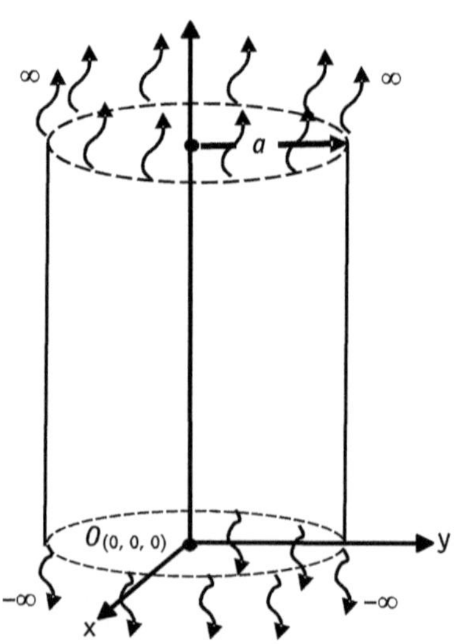

Figure 7.16 Free-space/dielectric material interface for Problem 7.14.

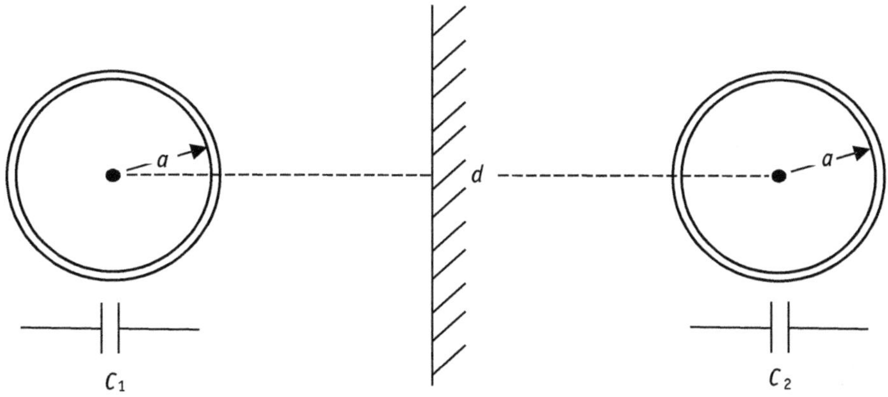

Figure 7.17 Free-space/dielectric material interface for Problem 7.15.

Problem 7.16

Find the capacitance of a parallel-plate capacitor containing two dielectrics, $\varepsilon_{r1} = 1.5$ and $\varepsilon_{r2} = 3.5$, each comprising one-half the total volume as shown in Figure 7.18.

Problem 7.17

Repeat the previous problem except for the parallel plates and two dielectrics, $\varepsilon_{r1} = 1.5$ and $\varepsilon_{r2} = 3.5$, as shown in Figure 7.19.

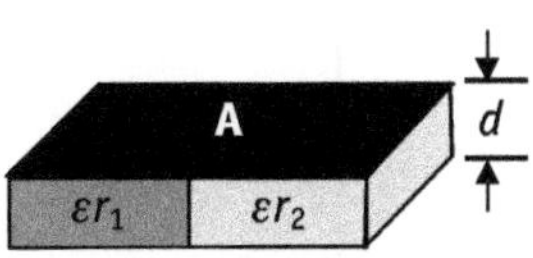

Figure 7.18 Free-space/dielectric material interface for Problem 7.16.

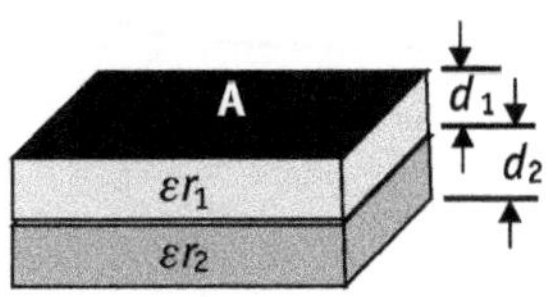

Figure 7.19 Free-space/dielectric material interface for Problem 7.17.

Problem 7.18

Each dielectric in the cylindrical capacitor (and there are two of them, see Figure 7.20) occupies one half of the volume. Find the capacitance.

$$C = 2\varepsilon_{r\text{-ave}}\,[(\pi\varepsilon_0 L)/\ln(b/a)] = [(2\pi\varepsilon_0 L)/\ln(b/a)]\varepsilon_{r\text{-ave}}$$

Problem 7.19

Find the voltage across each dielectric in the capacitor shown in Figure 7.21 when the applied voltage is 200V.

Problem 7.20

Find the voltage drop across each dielectric in Figure 7.22 where $\varepsilon_{r1} = 2.0$ and $\varepsilon_{r2} = 5.0$. The inner conductor is $r_1 = 2$ cm and the outer

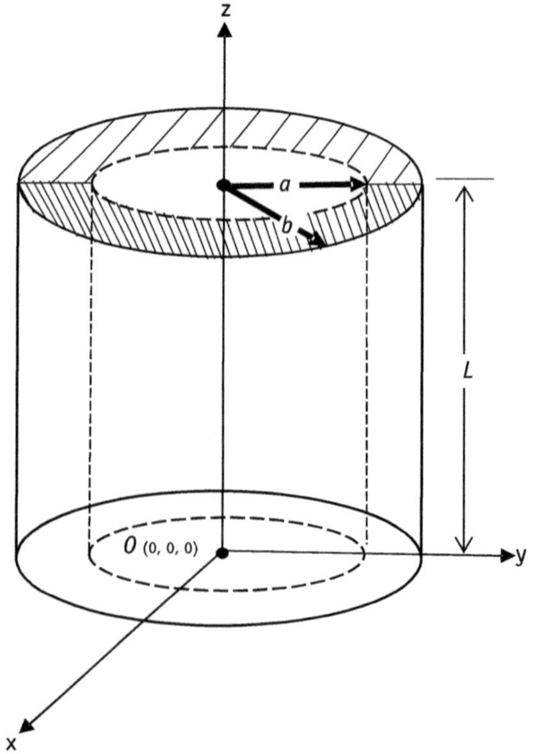

Figure 7.20 Free-space/dielectric material interface for Problem 7.18.

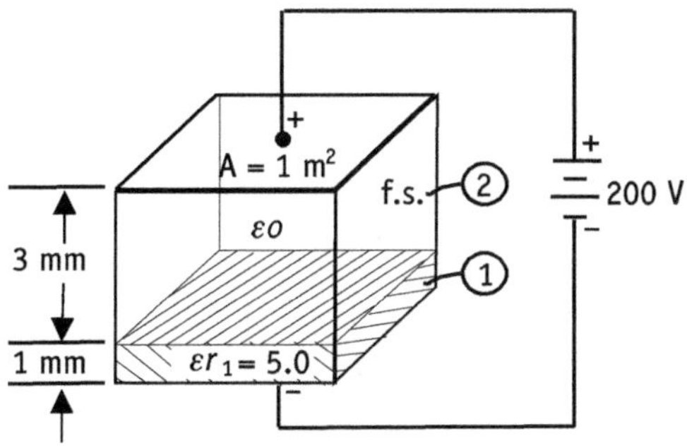

Figure 7.21 Free-space/dielectric material interface for Problem 7.19.

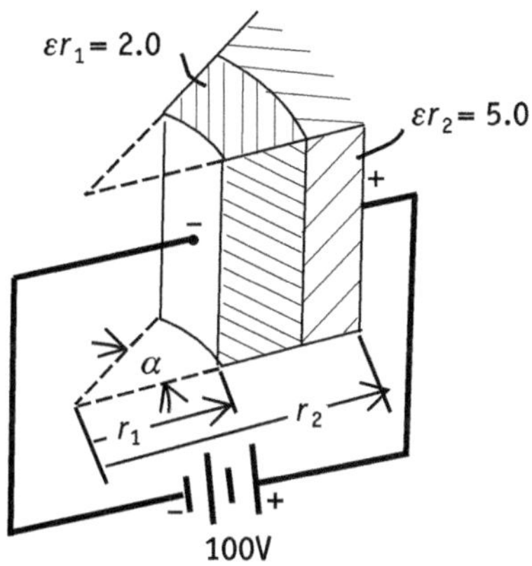

Figure 7.22 Free-space/dielectric material interface for Problem 7.20.

conductor is $r_2 = 2.5$ cm with dielectric interface that is located halfway between them.

References

[1] Russer, P., *Electromagnetics, Microwave Circuit and Antenna Design for Communications Engineering*, Norwood, MA: Artech House, 2006.

[2] Shadowitz, A., *The Electromagnetic Field*, New York: Dover Publications, Inc., 1975.

[3] Warnick, K. F., and P. Russer, *Problem Solving in Electromagnetics, Microwave Circuit and Antenna Design for Communications Engineering*, Norwood, MA: Artech House, 2006.

[4] Cheng, D. K., *Fundamentals of Engineering Electromagnetics*, 4th Edition, McGraw-Hill Education, 2019.

8

LAPLACE'S EQUATION

8.1 INTRODUCTION

Electric field intensity, **E**, was discussed in Chapter 2 and determined by summation or integration of point charges, line charges, and other charge configurations. In Chapter 3, Gauss' law was used to obtain **D**, which then gives **E**. While these two approaches are of value to an understanding of electromagnetic (EM) theory, they both tend to be impractical because charge distributions are not usually known. On the other hand, the method of Chapter 5, where **E** was found to be the negative of the gradient of V, requires that the potential function throughout the region be known. However, unfortunately, the potential function is generally not known.

Instead, conducting materials in the form of planes, curved surfaces, or lines are usually specified, and the voltage on one is also known with respect to some reference voltage that is located on one of the other conductors. Laplace's equation then provides a method whereby the potential function, V, can be obtained subject to the boundary conditions on the conductors [1].

8.2 POISSON'S EQUATION AND LAPLACE'S EQUATION

In Section 4.3, one of Maxwell's equations, $\nabla\mathbf{D} = \rho$, was developed. By substituting $\varepsilon\mathbf{E} = \mathbf{D}$ and $-\nabla V = \mathbf{E}$, we have:

$$\nabla \cdot \varepsilon(-\nabla V) = \rho \tag{8.1}$$

If the medium is homogeneous throughout the region of interest, ε may then be removed from the partial derivatives involved in the divergence. Hence, this gives:

$$\nabla \cdot \nabla V = -\frac{\rho}{\varepsilon} \quad \text{or} \quad \nabla^2 V = -\frac{\rho}{\varepsilon} \tag{8.2}$$

which is *Poisson's equation.* When the region of interest contains charges in a known distribution, ρ, Poisson's equation can be used to determine the potential function. However, the region very often is charge-free and has uniform permittivity. Poisson's equation then becomes *Laplace's equation* [2]:

$$\nabla^2 V = 0 \tag{8.3}$$

8.3 EXPLICIT FORMS OF LAPLACE'S EQUATION

Since the left side of Laplace's equation is the *divergence of the gradient of* V, these operations can be used to arrive at the form of the equation in a particular coordinate system. Mathematical displays of the divergence of the gradient of V are given, for each of the three coordinate systems (i.e., Cartesian, cylindrical, and spherical systems), as follows:

Cartesian coordinates:

$$\nabla V = \frac{\partial V}{\partial x}\mathbf{a}_x + \frac{\partial V}{\partial y}\mathbf{a}_y + \frac{\partial V}{\partial z}\mathbf{a}_z \tag{8.4}$$

For general vector field, $\mathbf{A}$,

$$\nabla \cdot \mathbf{A} = \frac{\partial A_x}{\partial x} + \frac{\partial A_y}{\partial y} + \frac{\partial A_z}{\partial z} \tag{8.5}$$

Hence, Laplace's equation in the Cartesian coordinate system, is [1]:

$$\nabla^2 V = \frac{\partial^2 V}{\partial x^2} + \frac{\partial^2 V}{\partial y^2} + \frac{\partial^2 V}{\partial z^2} = 0 \tag{8.6}$$

Cylindrical coordinates:

$$\nabla V = \frac{\partial V}{\partial r}\mathbf{a}_r + \left(\frac{1}{r}\right)\frac{\partial V}{\partial \varphi}\mathbf{a}_\varphi + \frac{\partial V}{\partial z}\mathbf{a}_z \tag{8.7}$$

For general vector field **A**,

$$\nabla \cdot \mathbf{A} = \left(\frac{1}{r}\right)\frac{\partial}{\partial x}(rA_r) + \left(\frac{1}{r}\right)\frac{\partial A_\varphi}{\partial \varphi} + \frac{\partial A_z}{\partial z} \tag{8.8}$$

Hence, Laplace's equation in the cylindrical coordinate system is [1]:

$$\nabla^2 V = \left(\frac{1}{r}\right)\frac{\partial}{\partial r}\left(\frac{r\,\partial V}{\partial \varphi}\right) + \left(\frac{1}{r^2}\right)\left(\frac{\partial^2 V}{\partial \varphi^2}\right) + \frac{\partial^2 V}{\partial z^2} = 0 \tag{8.9}$$

Spherical coordinates:

$$\nabla V = \frac{\partial V}{\partial r}\mathbf{a}_r + \left(\frac{1}{r}\right)\frac{\partial V}{\partial \varphi}\mathbf{a}_\varphi + \left(\frac{1}{r\sin\theta}\right)\frac{\partial V}{\partial \theta}\mathbf{a}_\theta \tag{8.10}$$

For a general vector field **A**,

$$\nabla \cdot \mathbf{A} = \left(\frac{2}{r^2}\right)\frac{\partial}{\partial r}(r^2 A_r) + \left(\frac{1}{r\sin\theta}\right)\frac{\partial}{\partial \phi}(A_\phi \sin\theta) + \left(\frac{1}{r\sin\theta}\right)\frac{\partial A_\phi}{\partial \phi} \tag{8.11}$$

Hence, Laplace's equation in the spherical coordinate system is [1]:

$$\nabla^2 V = \left(\frac{1}{r^2}\right)\frac{\partial}{\partial r}\left(\frac{r^2\,\partial V}{\partial r}\right) + \left(\frac{1}{r^2\sin\theta}\right)\frac{\partial}{\partial \theta}\left[\sin\theta\left(\frac{\partial V}{\partial \theta}\right)\right] + \left(\frac{1}{r^2\sin^2\theta}\right)\frac{\partial^2 V}{\partial \phi^2} = 0 \tag{8.12}$$

8.4 UNIQUENESS THEOREM

Any solution to either Laplace's or Poisson's equation that also satisfies the boundary conditions must be the only existing solution. Thus, that solution is said to be *unique*. Sometimes, there is confusion on this point. However, usually, that is due mainly to incomplete boundaries.

Consider the conducting plane at $z = 0$, shown in Figure 8.1, having a voltage of 100V.

It is clear that both:

$$V_1 = 5z + 100 \tag{8.13}$$

and

$$V_2 = 100 \tag{8.14}$$

satisfy Laplace's equation and the requirement that $V = 100$ when $z = 0$. The answer is that a single conducting surface with a specified voltage and no reference voltage given does not form the complete boundary of a properly defined region [3]. Even two finite parallel conducting planes do not form a complete boundary since the fringing of the field around the edges cannot be determined.

However, when parallel planes are specified and when it is stated to neglect fringing, then the region between the planes has boundaries that are properly defined.

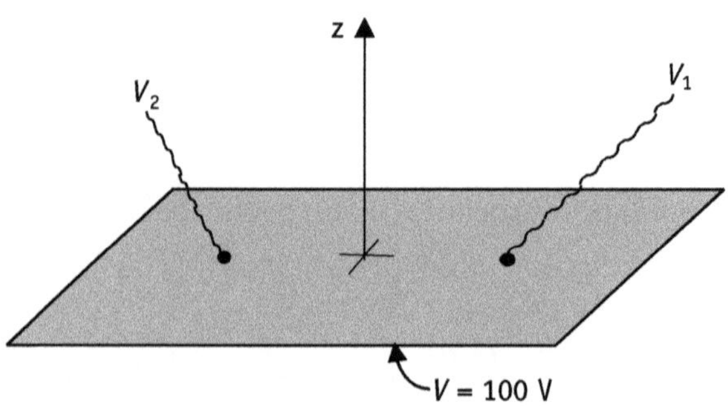

Figure 8.1 Conducting plane at $z = 0$.

8.5 MEAN VALUE AND MAXIMUM VALUE THEOREMS

Two important properties of the potential in a charge-free region can be obtained from Laplace's equation:

1. At the center of an included circle or sphere, the potential difference, V, is equal to the average of the values it assumes on the circle or sphere;
2. The potential difference, V, cannot have a maximum (or a minimum) within the region.

It then follows from the second above-stated property that any maximum of V must occur on the other boundary region. Since V obeys Laplace's equation:

$$\frac{\partial^2 V}{\partial x^2} + \frac{\partial^2 V}{\partial y^2} + \frac{\partial^2 V}{\partial z^2} = 0 \tag{8.15}$$

so do $\partial V/\partial x$, $\partial V/\partial y$, and $\partial V/\partial z$. Thus, *the components of the electric field intensity take their maximum values on the boundary* [4].

8.6 CARTESIAN SOLUTION IN ONE VARIABLE

Consider the parallel conductors of Figure 8.2.

In Figure 8.2, $V = 0$ at $z = 0$ and $V = 100$ at $z = d$. Assuming the region between the plates is charge-free,

$$\nabla^2 V = \frac{\partial^2 V}{\partial x^2} + \frac{\partial^2 V}{\partial y^2} + \frac{\partial^2 V}{\partial z^2} = 0 \tag{8.16}$$

With fringing neglected, the potential can vary only with z. Then,

$$\frac{d^2 V}{dx^2} = 0 \tag{8.17}$$

Next, integrating,

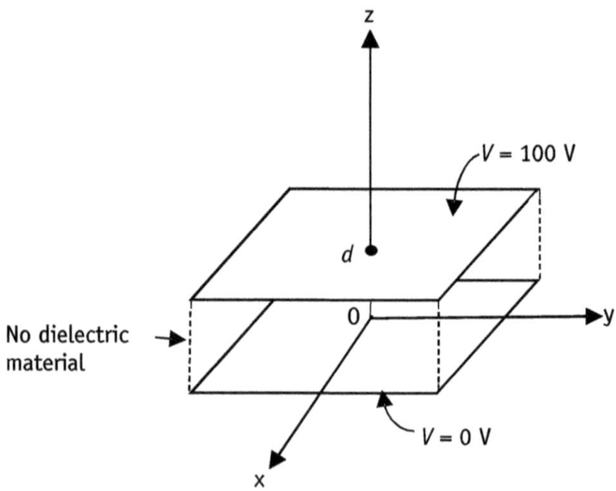

Figure 8.2 Parallel conductors (boundary conditions established).

$$V = Az + B \tag{8.18}$$

The boundary condition $V = 0$ at $z = 0$ requires that $B = 0$. $V = 100$ at $z = d$ gives $A = 100/d$.

Thus,

$$V = 100\left(\frac{z}{d}\right) \quad \text{(V)} \tag{8.19}$$

The electric field intensity, **E**, can now be obtained from:

$$\begin{aligned}\mathbf{E} &= -\nabla V = -\left(\frac{\partial V}{\partial x}\mathbf{a}_x + \frac{\partial V}{\partial y}\mathbf{a}_y + \frac{\partial V}{\partial z}\mathbf{a}_z\right) = -\frac{\partial}{\partial z}\left(100\frac{z}{d}\right)\mathbf{a}_z \\ &= \left(\frac{-100}{d}\right)\mathbf{a}_z\end{aligned} \tag{8.20}$$

Then,

$$\mathbf{D} = \varepsilon\left(\frac{-100}{d}\right)\mathbf{a}_z \quad \left(\text{C/m}^2\right) \tag{8.21}$$

At the conductors,

$$\rho_s = D_s = \pm\varepsilon\left(\frac{100}{d}\right) \quad (\text{C/m}^2) \tag{8.22}$$

where the plus sign applies at $z = d$ and the minus at $z = 0$ [2].

8.7 CARTESIAN PRODUCT SOLUTION

When the potential in Cartesian coordinates varies in more than one direction, Laplace's equation will contain more than one term. Suppose that V is a function of both x and y and has the special form $V = X(x)Y(y)$. Making use of the technique of separation of variables.

$$\frac{\partial^2(XY)}{\partial x^2} + \frac{\partial^2(XY)}{\partial y^2} = 0 \tag{8.23}$$

becomes

$$Y\frac{\partial^2 X}{\partial x^2} + X\frac{\partial^2(Y)}{\partial y^2} = 0 \tag{8.24}$$

or

$$\left(\frac{1}{X}\right)\frac{\partial^2 X}{\partial x^2} + \left(\frac{1}{Y}\right)\frac{\partial^2(Y)}{\partial y^2} = 0 \tag{8.25}$$

Since the first term is independent of y and the second term is independent of x, each may be set equal to a constant. However, the constant for one must be the negative of that for the other. Let the constant be a^2.

$$\left(\frac{1}{X}\right)\frac{\partial^2 X}{\partial x^2} = a^2 \tag{8.26}$$

$$\left(\frac{1}{Y}\right)\frac{\partial^2(Y)}{\partial y^2} = -a^2 \tag{8.27}$$

The general solution for X (for a given a) is:

$$X = A_1 e^{ax} + A_2 e^{-ax} \tag{8.28}$$

or, equivalently,

$$X = A_3 \cosh(ax) + A_4 \sinh(ax) \tag{8.29}$$

The general solution for Y (for a given a) is:

$$Y = B_1 e^{jay} + B_2 e^{-jay} \tag{8.30}$$

or, equivalently,

$$Y = B_3 \cos(ay) = B_4 \sin(ay) \tag{8.31}$$

Therefore, the potential function in the variables x and y can be written:

$$V = X(x)Y(y) = \left(A_1 e^{ax} + A_2 e^{-ax}\right)\left(B_1 e^{jay} + B_2 e^{-jay}\right) \tag{8.32}$$

where

$X(x) = (A_1 e^{ax} + A_2 e^{-ax})$
$Y(y) = (B_1 e^{jay} + B_2 e^{-jay})$

or

$$V = X(x)Y(y) = \left(A_3 \cosh ax + A_4 \sinh ax\right)\left(B_3 \cos ay + B_4 \sin ay\right) \tag{8.33}$$

where

$X(x) = (A_3 \cosh ax + A_4 \sinh ax)$
$Y(y) = (B_3 \cos ay + B_4 \sin ay)$

Because Laplace's equation is a linear, homogeneous equation, a sum of products of the above form is also a solution. Each product corresponds to a different value of a. The most general solution can be generated in this fashion.

Three-dimensional solutions, that is,

$$V = X(x)Y(y)Z(z) \tag{8.34}$$

of similar form to the above equation can be obtained. However, in this case there are two separation constants [4].

8.8 CYLINDRICAL PRODUCT SOLUTION

If a solution of the form,

$$V = R(r)\Phi(\Phi)Z(z) \tag{8.35}$$

is assumed, Laplace's equation becomes:

$$\left(\frac{\Phi Z}{r}\right)\frac{d}{dr}\left(r\frac{dR}{dr}\right)+\left(\frac{RZ}{r^2}\right)\frac{d^2\Phi}{d\Phi^2}+R\Phi\left(\frac{d^2Z}{dz^2}\right)=0 \tag{8.36}$$

Dividing by $R\Phi Z$ and expanding the r-derivative,

$$\left(\frac{1}{R}\right)\left(\frac{d^2R}{dr^2}\right)+\left(\frac{1}{Rr}\right)\left(\frac{dR}{dr}\right)+\left(\frac{1}{r^2\Phi}\right)\left(\frac{d^2\Phi}{d\Phi^2}\right)=-\frac{1}{Z}\left(\frac{d^2Z}{dz^2}\right)=-b^2 \tag{8.37}$$

The r and Φ terms contain no z, and the z term contains neither r nor Φ. They could be set equal to a constant, $-b^2$, as shown above. Then,

$$\frac{1}{Z}\left(\frac{d^2Z}{dz^2}\right)=b^2 \tag{8.38}$$

This equation was encountered in the Cartesian product solution. The solution is:

$$Z = C_1\cosh(bz)+C_2\sinh(bz) \tag{8.39}$$

Now the equation in r and Φ may be further separated as follows:

$$\left(\frac{r^2}{R}\right)\left(\frac{d^2R}{dr^2}\right)+\left(\frac{r}{R}\right)\left(\frac{dR}{dr}\right)+b^2r^2=-\left(\frac{1}{\Phi}\right)\left(\frac{d^2\Phi}{d\phi^2}\right)=a^2 \tag{8.40}$$

The resulting equation in Φ,

$$\left(\frac{1}{\Phi}\right)\left(\frac{d^2\Phi}{d\phi^2}\right)=-a^2 \tag{8.41}$$

has the solution

$$\Phi = C_3\cos(a\phi)+C_4\sin(a\phi) \tag{8.42}$$

The equation in r,

$$\left(\frac{d^2R}{dr^2}\right)+\left(\frac{1}{r}\right)\left(\frac{dR}{dr}\right)+\left[b^2-\left(\frac{a^2}{r^2}\right)\right]R=0 \tag{8.43}$$

is a form of Bessel's differential equation. Its solutions are in the form of power series called Bessel functions [1].

$$R = C_5J_a(br)+C_6N_a(br) \tag{8.44}$$

where

$$J_a(br)=\sum_{m=0}^{\infty}\frac{\left[(-1)^m(br/2)^{(a+2m)}\right]}{\left[m!\Gamma(a+m+1)\right]} \tag{8.45}$$

and

$$N_a(br)=\frac{\left[\cos(a\pi)J_a(br)-J_{-a}(br)\right]}{(\sin a\pi)} \tag{8.46}$$

The series $J_a(br)$ is known as a Bessel function of the *first kind, order a*.

If $a = n$, an integer, the gamma function in the power series may be replaced by $(n + m)!$, and $N_a(br)$ is a Bessel function of the *second kind, order a*. If $a = n$, an integer, $N_a(br)$ is defined as the limit of the above quotient as $a \rightarrow n$.

The function $N_a(br)$ behaves like $\ln r$ near $r = 0$ (see Figure 8.3). Therefore, it is not involved in the solution ($C_6 = 0$) whenever the potential is known to be finite at $r = 0$.

For integral order n and large argument x, the Bessel functions behave like damped sine waves:

$$J_n(x) \sim \left(\frac{2}{\pi x}\right)^{1/2} \cos\left[x - \left(\frac{\pi}{4}\right) - \left(\frac{n\pi}{2}\right)\right] \tag{8.47}$$

$$N_n(x) \sim \left(\frac{2}{\pi x}\right)^{1/2} \sin\left[x - \left(\frac{\pi}{4}\right) - \left(\frac{n\pi}{2}\right)\right] \tag{8.48}$$

See Figure 8.3.

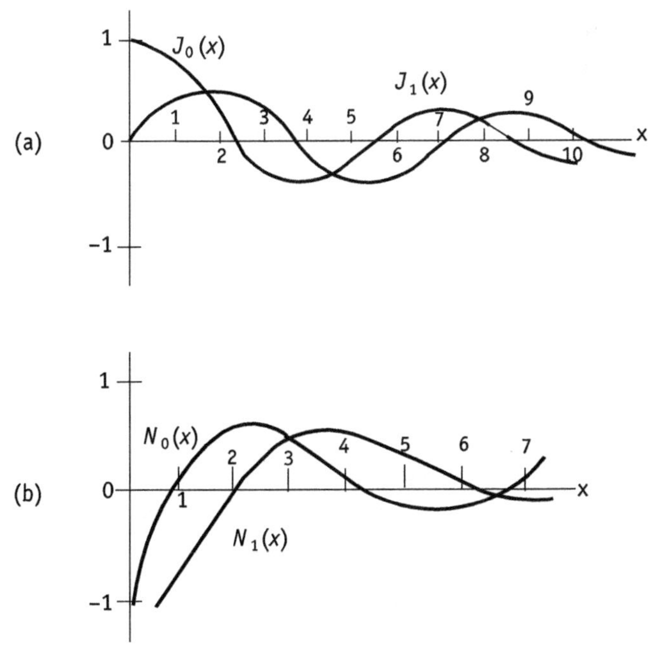

Figure 8.3 Bessel function (a) first kind, *n* order, (b) second kind, *n* order.

8.9 SPHERICAL PRODUCT SOLUTION

Of particular interest are those problems in spherical coordinates in which V may vary with r and θ but not with ϕ. For a product solution,

$$V = R(r)\Theta(\theta) \tag{8.49}$$

Laplace's equation becomes:

$$\left[\left(\frac{r^2}{R}\right)\left(\frac{d^2R}{dr^2}\right)+\left(\frac{2r}{R}\right)\left(\frac{dR}{dt}\right)\right]+\left[\left(\frac{1}{\Theta}\right)\left(\frac{d^2\Theta}{d\theta^2}\right)+\left\{\left[\frac{1}{(\Theta\tan\theta)}\right]\left(\frac{d\Theta}{d\theta}\right)\right\}=0\right. \tag{8.50}$$

The separation constant is chosen as $n(n + 1)$, where n is an integer (for reasons that will become apparent). The two separated equations are:

$$r^2\left(\frac{d^2R}{dr^2}\right)+2r\left(\frac{dR}{dr}\right)-n(n+1)R=0 \tag{8.51}$$

and

$$\left.\left(\frac{d^2\Theta}{d\theta^2}\right)+\left(\frac{1}{\tan\theta}\right)\left(\frac{d\Theta}{d\theta}\right)\right\}+n(n+1)\Theta=0 \tag{8.52}$$

The equation in r has the solution:

$$R = C_1 r^n + C_2 r^{-(n+1)} \tag{8.53}$$

The equation in θ possesses (unlike Bessel's equation) a polynomial solution of degree n in the variable $\xi = \cos\theta$, given by:

$$P_n(\xi)=\left(\frac{1}{2^n n!}\right)\left[\frac{d^n}{d\xi^n}(\xi^2-1)^n\right] \tag{8.54}$$

where $n = 0, 1, 2, \ldots$

The polynomial $P_n(\xi)$ is the *Legendre polynomial of order* n. There is a second, independent solution, $Q_n(\xi)$, which is logarithmically infinite at $\xi = \pm 1$ (i.e., $\theta = 0, \pi$) [4].

8.10 PROBLEMS AND SOLUTIONS: LAPLACE'S EQUATION

Problem 8.1

The potential has the value:

- V_1 on $1/n$ of the circle;
- 0 on the rest of the circle.

The entire region inside the circle is charge free (see Figure 8.4).

Find the potential at the center of the circle.

Problem 8.2

Show how mean value theorem follows the result of the previously solved problem.

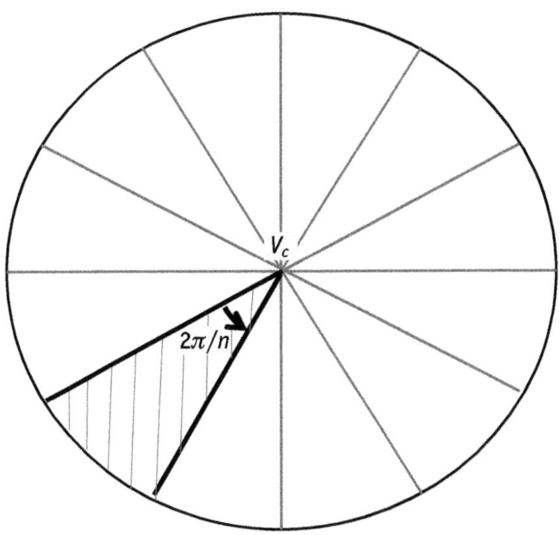

Figure 8.4 Circle highlighting a specific single sector.

Problem 8.3

Prove that with a charge-free region then potential cannot attain a maximum value.

Problem 8.4

Find the potential function for the region between the parallel circular disks of Figure 8.5. Neglect fringing.

Problem 8.5

Two parallel conducting plates in free space are at

$$y = 0 \quad \text{and} \quad y = 0.02\text{m}.$$

The zero-voltage reference is at $y = 0.01$m. If:

$$\mathbf{D} = 253\mathbf{y}\ (\text{nC/m}^2)$$

between the conductors, determine conductor voltages.

Problem 8.6

The parallel conducting disks in Figure 8.6 are separated by 5 mm and contain a dielectric for which $e_r = 2.2$.

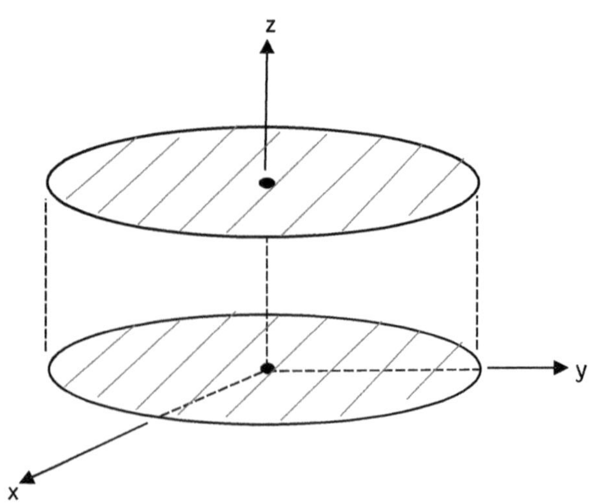

Figure 8.5 Circle divided into sectors.

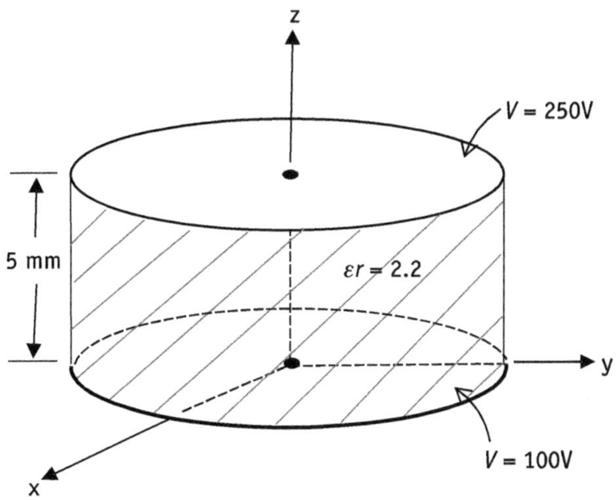

Figure 8.6 Circular disks separated by dielectric.

Determine charge densities on the disks.

References

[1] Boyce, W. E., *Calculus*, New York: John Wiley & Sons, 1988.

[2] Shadowitz, A., *The Electromagnetic Field*, New York: Dover Publications, 1975.

[3] Warnick, K. F., and P. Russer, *Problem Solving in Electromagnetics, Microwave Circuit and Antenna Design for Communications Engineering*, Norwood, MA: Artech House, 2006.

[4] Russer, P., *Electromagnetics, Microwave Circuit and Antenna Design for Communications Engineering*, Norwood, MA: Artech House, 2006.

PART 2

MAGNETOSTATICS: BASIC PRINCIPLES

9

AMPERE'S LAW AND THE MAGNETIC FIELD

9.1 MAGNETOSTATICS

A static magnetic field can originate from either a constant current source or a permanent magnet. This chapter is devoted to the discussion of nontime-varying magnetic fields that are produced from constant current sources. Time-variable magnetic fields, that is, those which coexist with time-variable electric fields, will be examined beginning in Chapter 13.

9.2 BIOT-SAVART LAW

A differential *magnetic field strength, dH,* results from a differential current element $I\,d\mathbf{l}$. The field varies inversely with distance squared. It is also independent of the surrounding medium and has a direction given by the cross-product of $I\,d\mathbf{l}$ and $\mathbf{a}_r$ (or $\mathbf{r}$) [1]. This relationship is known as the Biot-Savart law:

$$d\mathbf{H} = Id\mathbf{l} \times \frac{\mathbf{a}_r}{4\pi R^2} = Id\mathbf{l} \times \frac{r}{4\pi R^2} \quad \text{(A/m)} \tag{9.1}$$

The direction of $\mathbf{R}$ must be from the current element to the point at which $d\mathbf{H}$ is to be determined. This is shown in Figure 9.1.

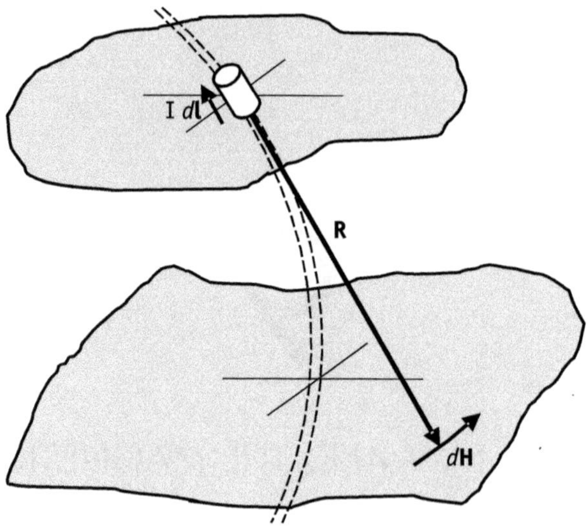

Figure 9.1 Generalized illustration of the Biot-Savart law.

Current elements have no separate existence. All elements making up the complete current filament contribute to **H** and, therefore, must be included. As a result, the summation leads to the integral form of the Biot-Savart law:

$$\mathbf{H}\oint I\, d\mathbf{l} \times \frac{\mathbf{a}_r}{4\pi R^2} = \oint I\, d\mathbf{l} \times \frac{r}{4\pi R^2} \quad \text{(A/m)} \qquad (9.2)$$

The closed line integral simply requires that all I current elements be included in order to obtain the complete **H** (the contour may close at ∞).

Example 9.1

An infinitely long, straight, filamentary current I along the z-axis in cylindrical coordinates is shown in Figure 9.2 [2].

A point in the $z = 0$ plane is selected with no loss in generality. In differential form, the equation may be expressed as:

$$d\mathrm{H} = I\, d z\, \mathbf{z} \times \frac{\left(r\mathbf{r} - z\mathbf{z}\right)}{4\pi\left(r^2 + z^2\right)^{3/2}}$$

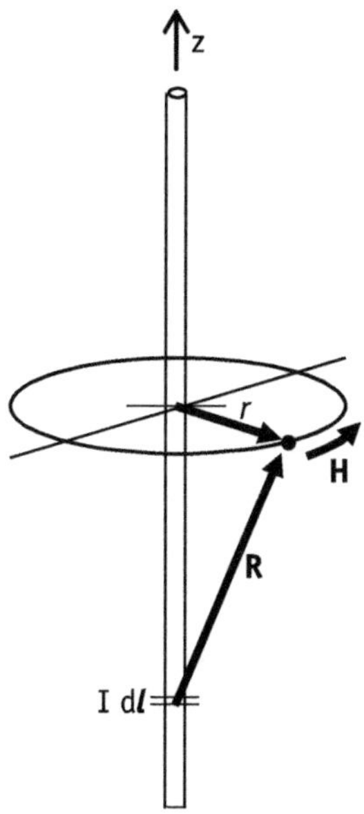

Figure 9.2 Biot-Savart law illustrated in cylindrical coordinates.

The variable of integration is z. Since $\mathbf{a}_\phi$ does not change with z, it may be removed from the integrand before integrating [3]. Hence,

$$\mathbf{H} = \left[\int_{-\infty}^{\infty} \frac{I\ r\ dz}{4\pi\left(r^2 + z^2\right)^{\frac{3}{2}}} \right] \mathbf{a}_\phi = \left(\frac{I}{2\pi r} \right) \mathbf{a}_\phi$$

This important result shows that **H** is inversely proportional to the radial distance, and the direction is in agreement with the right-hand rule; that is, where the four curled fingers of the right-hand point in the direction of the H-field when the z-axis aligned conductor is grasped by the hand such that the thumb points in the direction of the current (see Figure 9.2).

Continuing, the magnetic fields of sheet currents and volume currents are also given by the integral form of the Biot-Savart law. However, $I\,d\mathbf{l}$ is replaced by $\mathbf{K}dS$ (sheet current) or $\mathbf{J}dv$ (volume current) and with the integration extended, respectively, over the entire sheet or volume.

A particular case of importance is the infinite plane sheet of constant density, **K**. The field in this case is constant.

$$\mathbf{H} = \frac{1}{2}\mathbf{K} \times \mathbf{a}_n = \frac{1}{2}\mathbf{K} \times \mathbf{n} \quad \text{(A/m)} \tag{9.3}$$

where $\mathbf{n} = \mathbf{a}_n$.

9.3 AMPERE'S LAW

The line integral of the tangential component of **H** around a closed path is equal to the current enclosed by the path:

$$\oint \mathbf{H} \cdot d\mathbf{l} = I_{\text{enclosed}} \tag{9.4}$$

This is Ampere's law [1].

At first glance, a person would think that the law is used to determine the current I by integration. Instead, in practical cases, the current is usually known. The law provides a method of finding **H**. It is quite similar to the use of Gauss' law to find **D** given a known charge distribution.

In order to use Ampere's law to determine **H** there must be a considerable degree of symmetry in the problem. As such, two conditions must be met:

1. At each point of the closed path, **H** is either tangential or normal to the path;
2. H has the same value at all points of the path where **H** is tangential.

Thus, the Biot-Savart law can be used to help select a path which meets the two above-stated conditions [4]. In most cases, a proper path will be evident.

9.4 CURL

The curl of a vector field A results in another vector field. Point P in Figure 9.3 lies in the plane area ΔS bounded by the closed curve C.

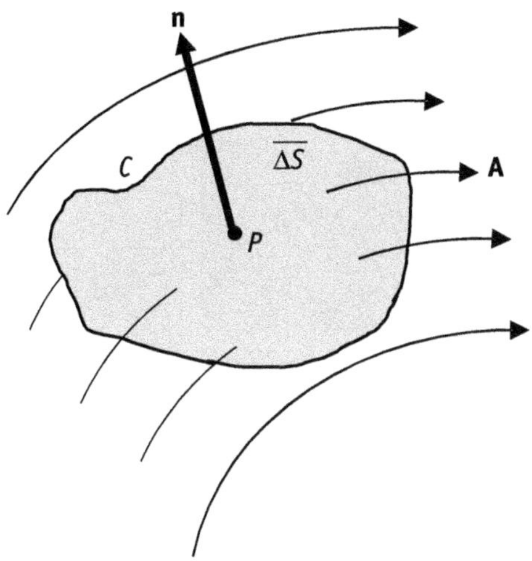

Figure 9.3 Illustration of the curl.

In the integration that defines the curl, C is traversed such that the enclosed area is as shown in the figure. The unit normal, **n**, (also sometimes denoted as $\mathbf{a}_n$ in vector mathematics), is determined by the right-hand rule (as shown in Figure 9.3). Then, the component of the curl of A in the direction **n**, is defined as [3]:

$$(\text{curl } \mathbf{A}) \cdot \mathbf{n} = \lim_{\Delta S \to 0} \frac{\left(\oint \mathbf{A} \cdot d\mathbf{l}\right)}{\Delta S} \tag{9.5}$$

In the coordinate systems, curl **A** is completely specified by its components along the three unit vectors. The x-component in Cartesian coordinates is defined by the taking as the contour C a square in the x = constant plane through P, as shown in Figure 9.4.

$$(\text{curl } \mathbf{A}) \cdot \mathbf{n}_x = \lim_{\Delta y \Delta x \to 0} \frac{\left(\oint \mathbf{A} \cdot d\mathbf{l}\right)}{\Delta y \Delta z} \tag{9.6}$$

If $\mathbf{A} = A_x\mathbf{a_x} + A_y\mathbf{a_y} + A_z\mathbf{a_z}$ at the corner of ΔS closest to the origin (point #1, see Figure 9.4), then,

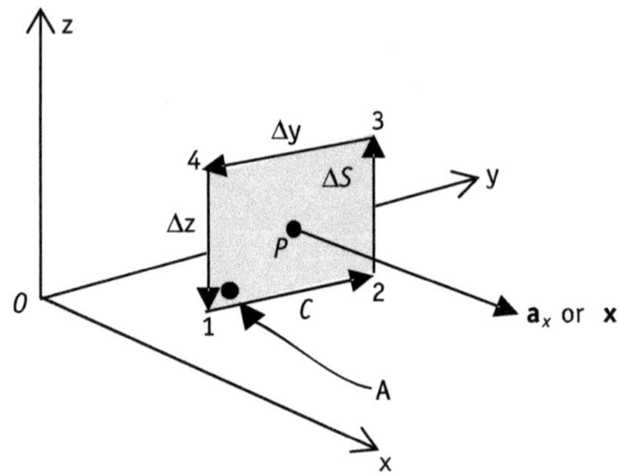

Figure 9.4 Curl in three components.

$$\oint = \int_1^2 + \int_2^3 + \int_3^4 + \int_4^1 = A_y\Delta_y + \left[A_x + \left(\frac{\partial A_z}{\partial y}\right)\Delta y\right]\Delta z$$
$$+\left[A_y + \left(\frac{\partial A_y}{\partial z}\right)\Delta z\right](-\Delta y) + A_x(-\Delta z) \tag{9.7}$$
$$= \left(\frac{\partial A_z}{\partial y} - \frac{\partial A_y}{\partial z}\right)\Delta y \Delta z$$

and,

$$(\text{curl } \mathbf{A}) \cdot \mathbf{a_x} = \left(\frac{\partial A_z}{\partial y} - \frac{\partial A_y}{\partial z}\right) \tag{9.8}$$

The y- and z-components can be determined in a similar fashion. Combining the three components:

Cartesian:

$$\text{curl } \mathbf{A} = \left(\frac{\partial A_z}{\partial y} - \frac{\partial A_y}{\partial z}\right)\mathbf{a_x} + \left(\frac{\partial A_x}{\partial z} - \frac{\partial A_z}{\partial x}\right)\mathbf{a_y} + \left(\frac{\partial A_y}{\partial x} - \frac{\partial A_x}{\partial y}\right)\mathbf{a_z} \tag{9.9}$$

A third-order determinant can be written. The expansion gives the Cartesian curl of **A**.

$$\text{curl } \mathbf{A} = \begin{vmatrix} \mathbf{a}_x & \mathbf{a}_y & \mathbf{a}_z \\ \partial/\partial x & \partial/\partial y & \partial/\partial z \\ A_x & A_y & A_z \end{vmatrix} \tag{9.10}$$

The elements in the second row are the del operator components (see Section 1.2). This suggests that $\nabla \times \mathbf{A}$ can be written for curl **A**. As with other expressions from vector analysis, this convenient notation is used for curl **A** in order coordinate systems even though ∇ is defined only in Cartesian coordinates.

Expressions for curl **A** in cylindrical and spherical coordinates can be derived in the same manner as above, though with more difficulty. See the following:

Cylindrical:

$$\begin{aligned} \text{curl } \mathbf{A} = {} & \left[\frac{1}{r}\frac{\partial A_z}{\partial \phi} - \frac{\partial A_\phi}{\partial z}\right]\mathbf{a}_r + \left(\frac{\partial A_r}{\partial z} - \frac{\partial A_z}{\partial r}\right)\mathbf{a}_\phi \\ & + \frac{1}{r}\left[\partial\left(\frac{rA_\theta}{\partial r} - \frac{\partial A_r}{\partial \phi}\right)\right]\mathbf{a}_z \end{aligned} \tag{9.11}$$

Spherical:

$$\begin{aligned} \text{curl } \mathbf{A} = {} & \frac{1}{(r\sin\theta)}\left[\frac{\partial(A_\phi \sin\theta)}{\partial\theta} - \frac{\partial A_\theta}{\partial\phi}\right]\mathbf{a}_r \\ & + \left(\frac{1}{r}\right)\left[\left(\frac{1}{\sin\theta}\right)\left(\frac{\partial A_r}{\partial\phi}\right) - \frac{\partial(rA_\phi)}{\partial r}\right]\mathbf{a}_\theta \\ & + \frac{1}{r}\left[\frac{\partial(rA_\theta)}{\partial r} - \frac{\partial A_r}{\partial\theta}\right)\mathbf{a}_\phi \end{aligned} \tag{9.12}$$

Two properties of the curl operator are frequently useful:

1. *The divergence of a curl is zero*; that is,

$$\nabla \cdot (\nabla \times \mathbf{A}) = 0 \tag{9.13}$$

for any vector field A;

2. *The curl of a gradient is zero*; that is,

$$\nabla \cdot (\nabla f) = 0 \tag{9.14}$$

for any scalar function of position f [5].

Under static conditions, $\mathbf{E} = -\nabla V$, and so from above property (2),

$$\nabla \times \mathbf{E} = 0 \tag{9.15}$$

9.5 CURRENT DENSITY *J* AND $\nabla \times \mathbf{H}$

The x-component of $\nabla \times \mathbf{H}$ is determined by $\oint \mathbf{H} \cdot d\mathbf{l}$, where the path lies in a plane normal to the x-axis. Ampere's law states that this integral is equal to the current enclosed. The direction is $\mathbf{a_x}$ so the current can be called I_x [1]. Thus,

$$(\text{curl } \mathbf{A}) \cdot \mathbf{a_x} = \lim_{\Delta S \to 0} \frac{I_x}{\Delta S} = J_x \tag{9.16}$$

The x-component of the current density $\mathbf{J}$. Likewise for the y and z directions. Consequently,

$$\nabla \times \mathbf{H} = \mathbf{J} \tag{9.17}$$

This important result is one of Maxwell's equations for static fields. If $\mathbf{H}$ is known throughout a particular region, then $\nabla \times \mathbf{H}$ will produce $\mathbf{J}$ for that region.

9.6 MAGNETIC FLUX DENSITY B

Like $\mathbf{D}$, the magnetic field strength $\mathbf{H}$ depends only on (moving) charges and is independent of the medium. The force field associated with $\mathbf{H}$ is the *magnetic flux density* $\mathbf{B}$, which is given by

$$\mathbf{B} = \mu \mathbf{H} \tag{9.18}$$

where: $\mu = \mu_0\mu_r$ is the *permeability* of the medium. The unit of **B** is the *tesla*,

$$1\text{T} = 1\ \text{N/Am} \tag{9.19}$$

The free-space permeability μ_0 has the numerical value of $4\pi \times 10^{-7}$ and has the units henries per meter, H/m; μ_r, the *relative permeability* of the medium, is a pure number very near to unity except for a small group of ferromagnetic materials which will be treated in Chapter 11.

Magnetic flux, Φ, through a surface is defined as:

$$\Phi = \int_s \mathbf{B} \cdot d\mathbf{S} \tag{9.20}$$

The sign on Φ may be positive or negative depending upon the choice of the surface normal in $d\mathbf{S}$. The unit of magnetic flux is the weber, Wb [4]. The various magnetic units are related by:

$$1\text{T} = 1\ \text{Wb/m}^2 \tag{9.21}$$

$$1\text{H} = 1\ \text{Wb}/A \tag{9.22}$$

Example 9.2

Find the flux crossing the portion of the plane $\phi = \pi/4$ defined by $0.01\text{m} < r < 0.05\text{m}$ and $0 < z < 2\text{m}$ (see Figure 9.5). A current filament carrying 2.50A of current along the z-axis is in the $\mathbf{a}_z$ or z direction.

Solution: See the following equations:

$$\mathbf{B} = \mu_0\mathbf{H} = \mu_0\left(\frac{I}{2\pi r}\right)\phi_\phi = \mu_0\left(\frac{I}{2\pi r}\right)\phi$$

$$dS = dr\,dz\,\mathbf{a}_\phi = dr\,dz\,\phi$$

$$\mathbf{\Phi} = \int_0^2\int_{0.01}^{0.05}\left(\frac{\mu_0 I}{2\pi r}\right)\phi\cdot(dr\,dz))\phi = \left(\frac{2\mu_0 I}{2\pi}\right)\ln\left(\frac{0.05}{0.01}\right)$$

$$= 1.61 \times 10^{-6}\ \text{Wb or } 1.61\ \mu\text{Wb}$$

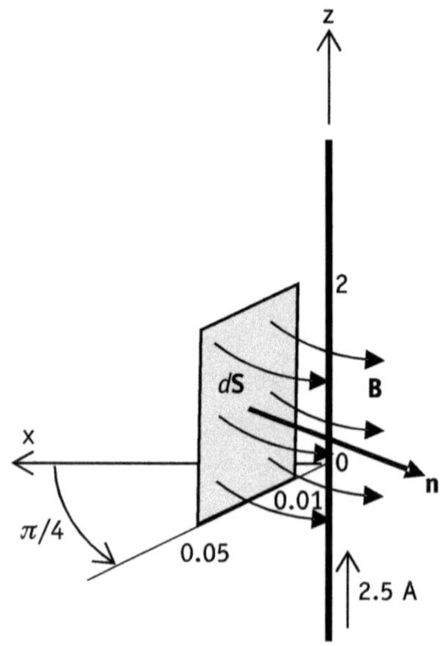

Figure 9.5 Magnetic flux crossing a portion of a plane.

It should be observed that the lines of magnetic flux Φ are closed curves with no starting or termination points. This is in contrast with electric flux ψ which originates on positive charge and terminates on negative charge. In Figure 9.6, all of the magnetic flux Φ that enters the closed surface must leave the surface.

Thus, **B** fields have no sources or sinks which is mathematically expressed by:

$$\nabla \cdot \mathbf{B} = 0 \tag{9.23}$$

9.7 VECTOR MAGNETIC POTENTIAL A

Electric field intensity **E** was first obtained from known charge configurations. Later, electric potential V was developed, and it was found that **E** could be obtained as the *negative gradient of* V; that is, $\mathbf{E} = -\nabla V$. Laplace's equation provided a method of obtaining V from known potentials on the boundary conductors. Similarly, a *vector magnetic potential*, **A**, is defined such that

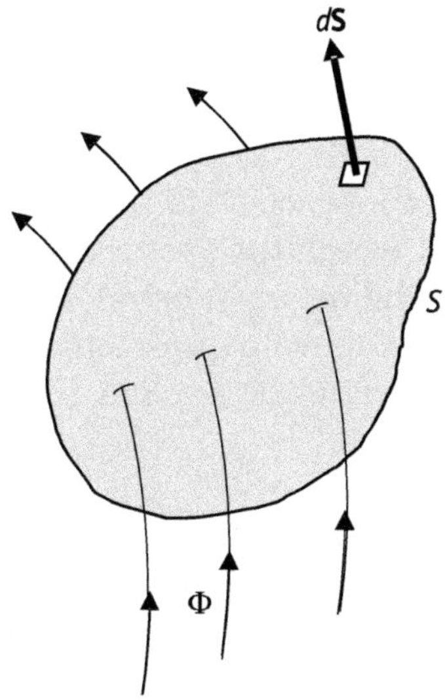

Figure 9.6 Magnetic flux entering the closed surface must exit the surface.

$$\nabla \times \mathbf{A} = \mathbf{B} \tag{9.24}$$

It serves as an intermediate quantity from which **B**, and, hence, **H** can be calculated. Note that the definition of **A** is consistent with the requirement that $\nabla \cdot \mathbf{B} = 0$. The units of **A** are Wb/m or T · m.

If the additional condition:

$$\nabla \cdot \mathbf{A} = 0 \tag{9.25}$$

is imposed, then the vector magnetic potential **A** can be determined form the known currents in the region of interest. For the three standard current configurations, the expressions are:

$$\text{Current filament:} \quad \mathbf{A} = \oint \left(\frac{\mu I}{4\pi r} \right) d\mathbf{l} \tag{9.26}$$

$$\text{Sheet current:} \quad \mathbf{A} = \int_S \left(\frac{\mu \mathbf{K}}{4\pi R} \right) dS \tag{9.27}$$

$$\text{Volume current:} \quad \mathbf{A} = \int_V \left(\frac{\mu \mathbf{J}}{4\pi R} \right) dv \tag{9.28}$$

R in each of the above cases is the distance from the current element to the point at which the vector magnetic potential is being calculated. Like the analogous integral for the electric potential (review Section 5.4), the above expressions for **A** presuppose a zero level at infinity. They cannot be applied if the current distribution itself extends to infinity [4].

Example 9.3

Investigate the vector magnetic potential for the infinite, straight, current filament I in free space. In Figure 9.7, the current filament is along the z-axis. The observation point is (x, y, z).

The particular current element

$$I\,d\mathbf{l} = I\,dl\,\mathbf{a}_z = I\,dl\,\mathbf{z}$$

at $l = 0$ is shown, where l is the running variable along the z-axis.

It is clear that the integral

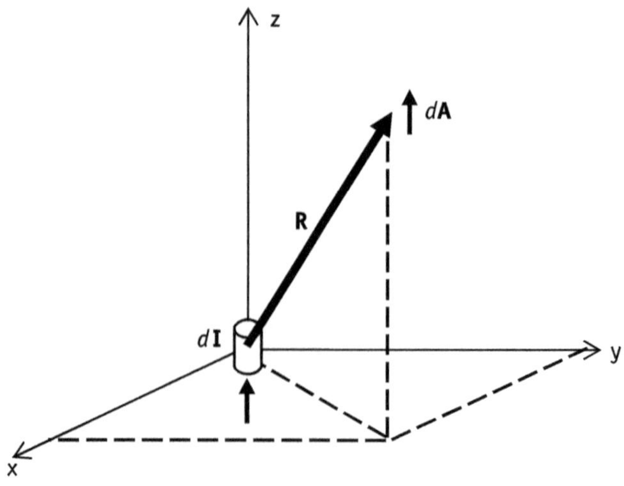

Figure 9.7 Vector magnetic potential illustration.

$$\mathbf{A} = \int_{-\infty}^{+\infty} \frac{(\mu_0 I\, dl))}{(4\pi R)} \mathbf{z}$$

does *not* exist since, when l is large, $\mathbf{R} \simeq l$. This is a case of a current distribution that extends to infinity.

However, it is possible to consider the *differential* vector potential:

$$d\mathbf{A} = \left\{ \frac{(\mu_0 I\, dl)}{\left[4\pi\left(x^2 + y^2 + z^2\right)^{1/2}\right]} \right\} \mathbf{z}$$

and

$$d\mathbf{B} = \nabla \times d\mathbf{A}$$

$$= \left(\frac{\mu_0 I\, dl}{4\pi}\right)\left[\left[\frac{-y}{\left(x^2 + y^2 + z^2\right)^{3/2}}\right]\mathbf{x} + \left[\frac{x}{\left(x^2 + y^2 + z^2\right)^{3/2}}\right]\mathbf{y}\right]$$

This result agrees with that for $d\mathbf{H} = (1/\mu_0)\, d\mathbf{B}$ as given by the Biot-Savart law.

9.8 STOKES' THEOREM

Consider an open surface S in which the boundary is a closed curve C. *Stokes' theorem* states that the integral of the tangential component of a vector field **F** around C is equal to the integral of the normal component of curl **F** over S [3].

$$\oint \mathbf{F} \cdot d\mathbf{l} = \int_S (\nabla \times \mathbf{F}) \cdot d\mathbf{S} \tag{9.29}$$

If **F** is chosen to be the vector magnetic potential **A**, Stokes' theorem gives:

$$\oint \mathbf{A} \cdot d\mathbf{l} = \int_S \mathbf{B} \cdot d\mathbf{S} = \Phi \tag{9.30}$$

9.9 PROBLEMS AND SOLUTIONS

Problem 9.1

An infinitely long, straight, filamentary current, I, runs along the z-axis as shown in Figure 9.8. In cylindrical coordinates, derive the equation for the H-field as a function of the current and the distance, r, away from the filament.

Problem 9.2

In this problem, use Ampere's law to obtain **H** for the same figure (Figure 9.8) of the previous problem.

Problem 9.3

Find the flux crossing the portion of the plane $\phi = \pi/4$ defined by $0.01\text{m} < r < 0.05\text{m}$ and $0\text{m} < z < 2\text{m}$. See Figure 9.9. A current filament of 2.5A along the z-axis is in the z direction.

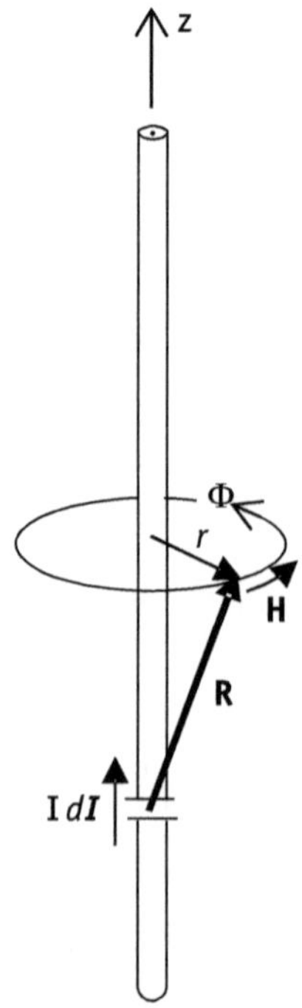

Figure 9.8 Biot-Savart law (i.e., H-field created from induced current, I).

Problem 9.4

Perform an investigation of the vector magnetic potential for the infinite, straight current filament, *I*, in free space (see Figure 9.10).

Problem 9.5

Find **H** at the center of the square current loop having sides of length *L* (see Figure 9.11).

Problem 9.6

A current filament of 5.0A in the *y* direction is parallel to the *y*-axis at $x = 2\text{m}$, $z = -2\text{m}$. Find **H** at the origin (see Figure 9.12).

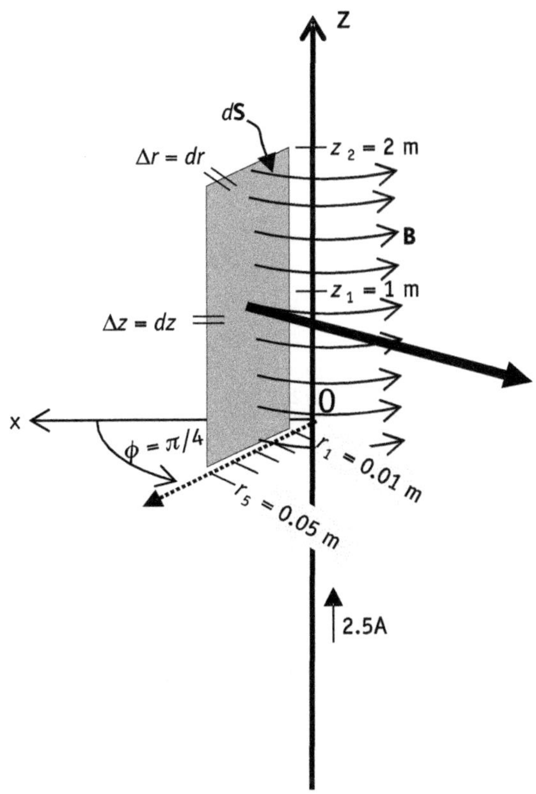

Figure 9.9 Ampere's law (i.e., current created from induced H-field).

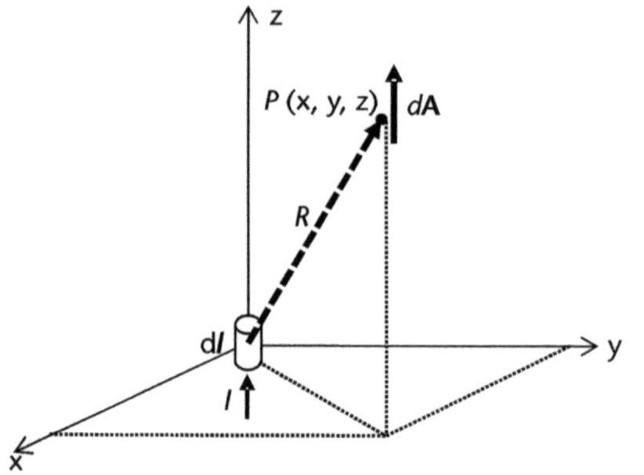

Figure 9.10 Illustration for vector magnetic potential investigation.

Problem 9.7

Determine an expression for **H** that is due to an infinite plane current sheet of uniform density, **K** (see Figure 9.13).

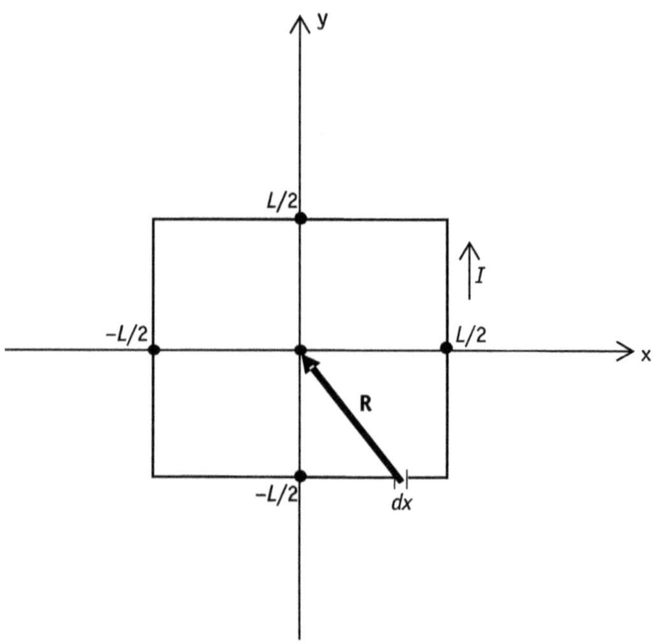

Figure 9.11 Illustration for Problem 9.5.

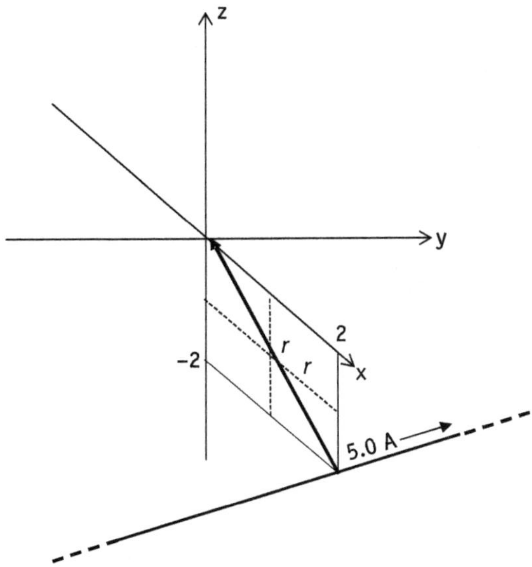

Figure 9.12 Illustration for Problem 9.6.

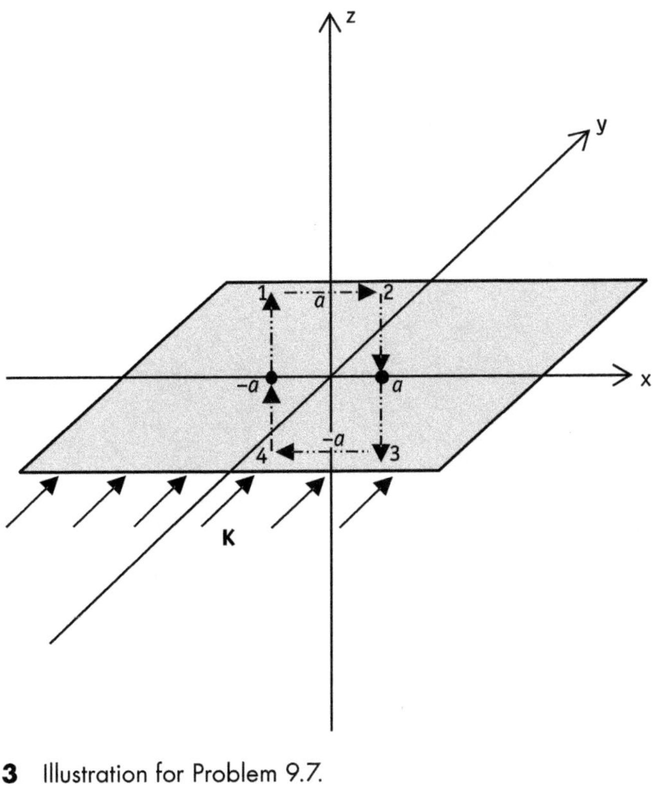

Figure 9.13 Illustration for Problem 9.7.

Problem 9.8

A current sheet, $K = 10\mathbf{z}$ (A/m), lies in the $x = 5$m plane. A second sheet, $\mathbf{K} = -10\mathbf{z}$ (A/m), lies in the $x = -5$m plane. Find $\mathbf{H}$ at all points (see Figure 9.14).

Problem 9.9

A thin hollow cylinder conductor of radius a and infinite length carries a current, $\mathbf{I}$. Find H at all points using Ampere's law (see Figure 9.15).

Problem 9.10

Determine $\mathbf{H}$ for a solid cylindrical conductor of radius a where the current, $\mathbf{I}$, is uniformly distributed over the cross section (see Figure 9.16).

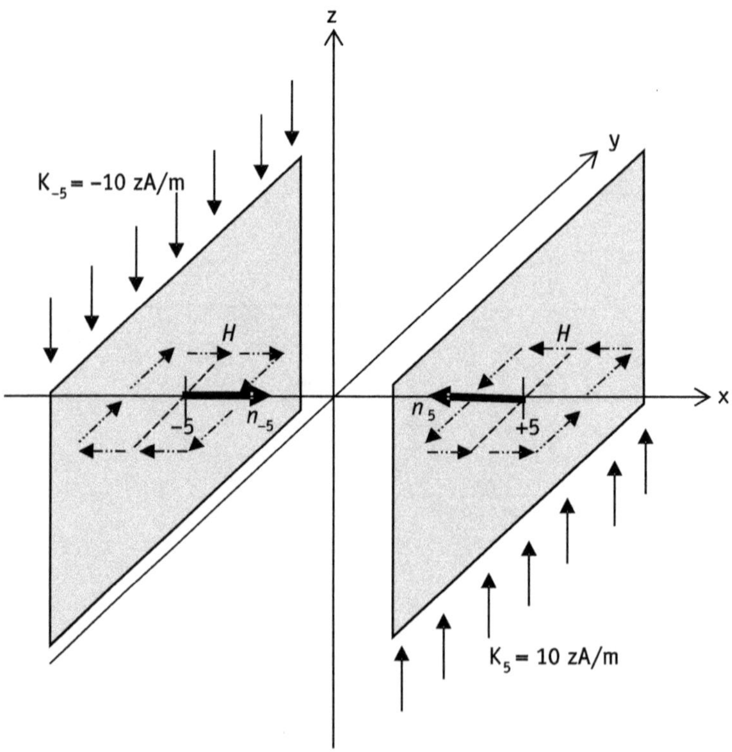

Figure 9.14 Illustration for Problem 9.8.

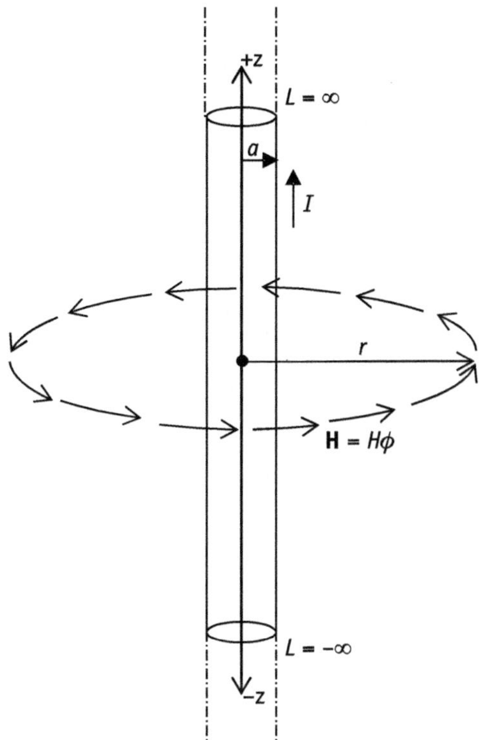

Figure 9.15 Illustration for Problem 9.9.

Problem 9.11

In the region $0 < r < 0.5$m in cylindrical coordinates, the current density is

$\mathbf{J} = 4.5\ e^{-2r}\mathbf{z}$

and

$\mathbf{J} = 0\ \mathbf{z}$ elsewhere.

Use Ampere's law to find **H** (see Figure 9.17).

Problem 9.12

A current sheet, $\mathbf{K} = 6.0\mathbf{x}$ (A/m), lies in the $z = 0$ plane with a current located $y = 0$, $z = 4$m shown in Figure 9.18.

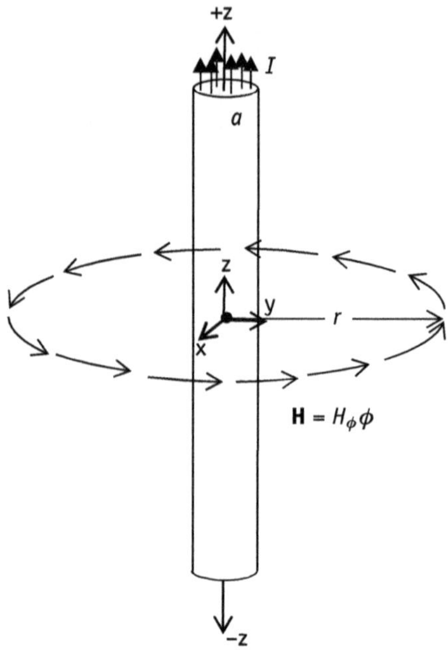

Figure 9.16 Illustration for Problem 9.10.

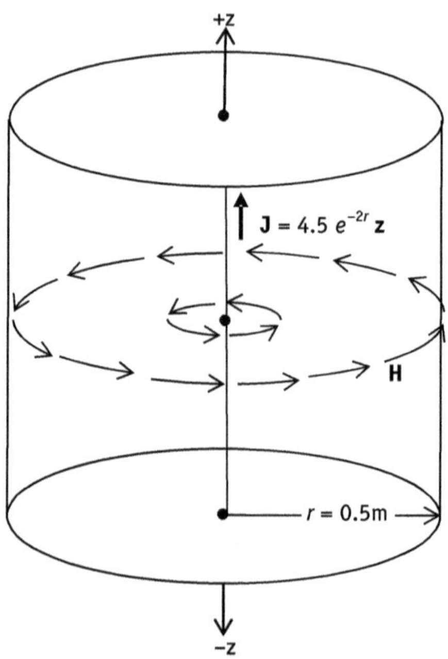

Figure 9.17 Illustration for Problem 9.11.

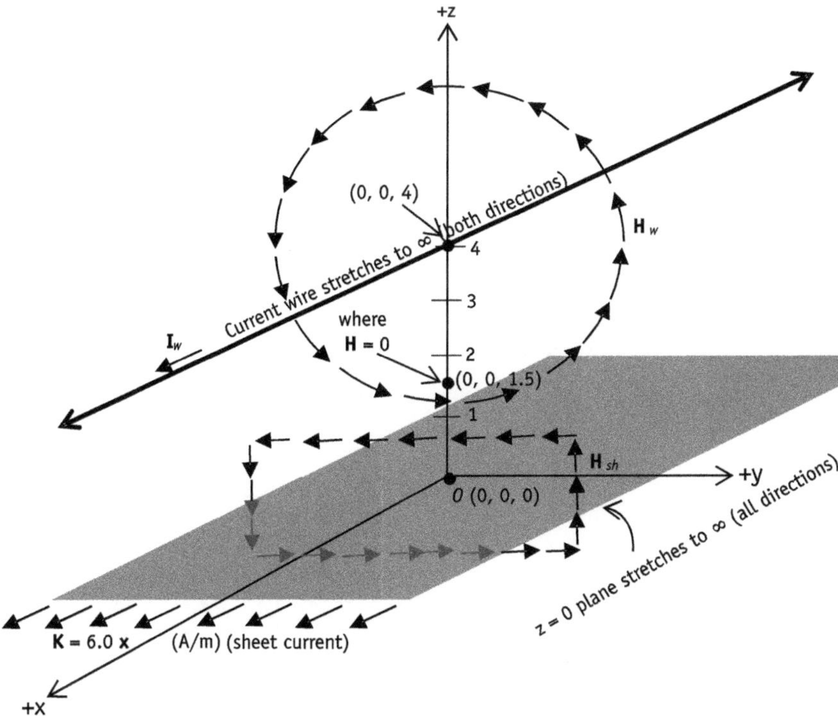

Figure 9.18 Illustration for Problem 9.12.

Determine I_w and its direction of $\mathbf{H} = 0$ at point (0, 0, 1.5m).

References

[1] Shadowitz, A., *The Electromagnetic Field*, New York: Dover Publications, Inc., 1975.

[2] Warnick, K. F., and P. Russer, *Problem Solving in Electromagnetics, Microwave Circuit and Antenna Design for Communications Engineering*, Norwood, MA: Artech House, 2006.

[3] Boyce, W. E., *Calculus*, New York: John Wiley & Sons, 1988.

[4] Russer, P., *Electromagnetics, Microwave Circuit and Antenna Design for Communications Engineering*, Norwood, MA: Artech House, 2006.

[5] Cheng, D. K., *Fundamentals of Engineering Electromagnetics*, Reading, MA: Addison-Wesley Publishing Company, Inc., 1993.

10

FORCES AND TORQUES IN MAGNETIC FIELDS

10.1 MAGNETIC FORCE ON PARTICLES

A charged particle *in motion* in a magnetic field experiences a force at right angles to its velocity—with a magnitude proportional to the charge—the velocity and the magnetic flux density. The complete expression is given by the cross-product [1].

$$\mathbf{F} = Q\mathbf{U} \times \mathbf{B} \tag{10.1}$$

Therefore, the direction of a particle in motion can be changed by a magnetic field. The magnitude of the velocity, U, and, consequently, the kinetic energy, will remain the same. This is in contrast to the electric field where the force

$$\mathbf{F} = Q\mathbf{E} \tag{10.2}$$

does work on the particle and, for this reason, changes its kinetic energy [1].

If the field $\mathbf{B}$ is uniform throughout a region and the particle has an initial velocity normal to the field, the path of the particle is a circle of a certain radius, r. The force of the field is of magnitude

$$F = |Q|UB \tag{10.3}$$

and is directed toward the center of the circle. The centripetal acceleration is of magnitude

$$\omega^2 r = \frac{U^2}{r} \tag{10.4}$$

Then, by Newton's second law:

$$|Q|UB = m\left(\frac{U^2}{r}\right) \tag{10.5}$$

or

$$r = \frac{mU}{|Q|B} \tag{10.6}$$

Observe that r is a measure of the particle's linear momentum, mU [2]. See Examples 10.1 and 10.2.

Example 10.1

Find the force on a particle of mass 1.70×10^{-17} kg and charge 1.60×10^{-19} C if it enters a field $\mathbf{B} = 5$ mT with an initial speed of 83.5 km/s.

Solution: See Figure 10.1. Unless directions are known for $\mathbf{B}$ and $\mathbf{U}_o$, the particle's initial velocity, the force cannot be calculated. Therefore, under the assumption that $\mathbf{B}$ and $\mathbf{U}_o$ are perpendicular to each other, the following may be performed:

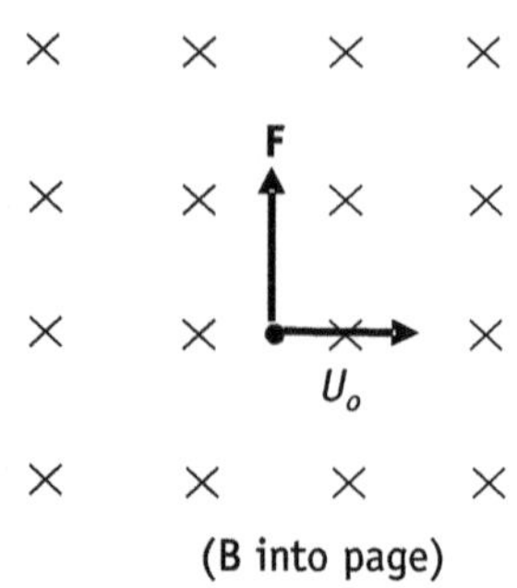

Figure 10.1 Magnetic force illustration for Example 10.1.

$$F = |Q|UB = |1.60 \times 10^{-19}\ \text{C}|(83.5 \times 10^{3}\ \text{m/s})(5 \times 10^{-3}\ \text{T})$$
$$= 6.68 \times 10^{-17}\ \text{N}$$

where

$Q = 1.60 \times 10^{-19}$ C

$U = 83.5 \times 10^{3}$ m/3

$B = 5$ mT

Example 10.2

For the Example 10.1 particle, find the radius of the circular path and the time required for one revolution.

$$r = \frac{mU}{|Q|B} = \frac{(1.70 \times 10^{-17}\ \text{kg})(83.5 \times 10^{3}\ \text{m/s})}{(1.60 \times 10^{-19}\ \text{C})(5 \times 10^{-3}\ \text{T})}$$
$$= 0.1774\ \text{m}$$

where

$m = 1.70 \times 10^{-17}$ kg

$U = 83.5 \times 103$ m/s

$Q = 1.60 \times 10^{-19}$ C

$B = 5$ mT

$$T = \frac{2\pi r}{U} = 2\pi\frac{(0.1774\ \text{m})}{(83.5 \times 10^{3}\ \text{m/s})}$$
$$= 13.5177\ \mu\text{s}$$

where

$r = 0.1774$m

$U = 83.5 \times 103$ m/s

10.2 ELECTRIC AND MAGNETIC FIELDS COMBINED

When both fields are present in a region at the time, the force on a particle is given by:

$$\mathbf{F} = Q(\mathbf{E} + \mathbf{U} \times \mathbf{B}) \tag{10.7}$$

This force, together with the initial conditions, determines the path of the particle (see Example 10.3) [1].

Example 10.3

A region contains a magnetic flux density $\mathbf{B} = 5.0 \times 10^{-4}\ \mathbf{a_z}$ T and an electric field $\mathbf{E} = 5.0\ \mathbf{a_z}$ V/m. A proton ($Q_p = 1.602 \times 10^{-19}$ C, $m_p = 1.673 \times 10^{-27}$ kg) enters the fields at the origin with an initial velocity $\mathbf{U_o} = 2.5 \times 15^5\ \mathbf{a_x}$ m/s. Describe the proton's motion and give its position after three complete revolutions.

Solution: The initial force on the particle is:

$$\mathbf{F} = Q\left(\mathbf{E} + \mathbf{U}_0 \times \mathbf{B}\right) = Q_p\left(E\mathbf{a}_z - U_0\mathbf{a}_y\right)$$

The z-component (electric component) of the force is constant and produces a constant acceleration in the z-direction. Thus, the equation of motion in the z-direction is:

$$z = \frac{1}{2}at^2 = \frac{1}{2}\left(\frac{Q_pE}{m_p}\right)t^2$$

The other magnetic component, which changes into $-Q_pU\,B\mathbf{a}_r$, produces circular motion that is perpendicular to the z-axis with period equal to:

$$T = \frac{2\pi r}{U} = \frac{2\pi m_p}{Q_pB}$$

The resultant motion is helical as shown in Figure 10.2.

After three revolutions, $x = y = 0$ and

$$z = \frac{1}{2}\left(\frac{Q_pE}{m_p}\right)(3T)^2 = \frac{18\pi^2 Em_p}{Q_pB^2} = 37.0\text{m}$$

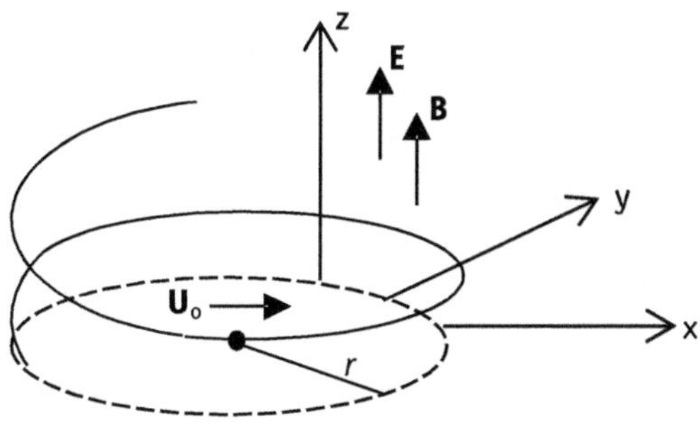

Figure 10.2 Illustration of electric and magnetic fields combined.

10.3 MAGNETIC FORCE ON A CURRENT ELEMENT

A frequently encountered situation is that of a current-carrying conductor in an external magnetic field. Since

$$I = \frac{dQ}{dt} \tag{10.8}$$

the different force equation may be written

$$d\mathbf{F} = dQ(\mathbf{U} \times \mathbf{B}) = (I\,dt)(\mathbf{U} \times \mathbf{B}) = I(d\mathbf{I} \times \mathbf{B}) \tag{10.9}$$

where:

$$d\mathbf{l} = \mathbf{U}dt \tag{10.10}$$

is the elementary length in the direction of the conventional current *I*. If the conductor is straight and the field is constant along it, the differential force may be integrated to give:

$$F = ILB\sin\theta \tag{10.11}$$

The magnetic force is in fact exerted on the electrons that make up the current *I*. However, since the electrons are confined to the conductor, the force is effectively transferred to the heavy lattice. This transferred force does the work on the whole conductor. While this fact provides a reasonable introduction to the behavior of current-carrying conductors in electric machines, certain essential considerations have been omitted. No mention was made, nor will be made in Section 10.4, of the current source and the energy that would be required to maintain a constant current *I*. Faraday's law of induction (see Section 12.3) was not applied. In electric machine theory, the result will be modified by these considerations [2].

Conductors in motion in magnetic fields are treated again in Chapter 12.

Example 10.4

Find a force on a straight conductor of length 0.30m carrying a current of 5.0A in the $-\mathbf{a}_z$ direction, where a field is $\mathbf{B} = 3.5 \times 10^{-3}$ $(\mathbf{a}_x - \mathbf{a}_y)$ T.

Solution: See the expressions given below:

$$\mathbf{F} = I(\mathbf{L} \times \mathbf{B}) = I\left[L(-\mathbf{a}_z) \times B(\mathbf{a}_x - \mathbf{a}_y)\right]$$

$$= (5.0\ \text{A})\left\{\left[(0.30\ \text{m})(-\mathbf{a}_z)\right] \times \left[(3.5 \times 10^{-3}\ \text{T})(\mathbf{a}_x - \mathbf{a}_y)\right]\right\}$$

$$= 7.42 \times 10^{-3}\left[\frac{(-\mathbf{a}_x - \mathbf{a}_y)}{(2^{1/2})}\right]\text{N}$$

where

$I = 5.0\text{A}$

$\mathbf{L} = L(-\mathbf{a}_z) = 0.30\text{m}\ (-\mathbf{a}_z)$

$L = 0.30\text{m}$

$\mathbf{B} = B(\mathbf{a}_x - \mathbf{a}_y) = 3.5 \times 10^{-3}\ \text{T}\ (\mathbf{a}_x - \mathbf{a}_y)$

$B = 3.5 \times 10^{-3}\ \text{T}$

The force having a magnitude of 7.42 mN is at right angles to both the B-field and the current direction as shown in Figure 10.3.

10.4 WORK AND POWER

The magnetic forces on the charged particles and current-carrying conductors examined above result from the field. To counter

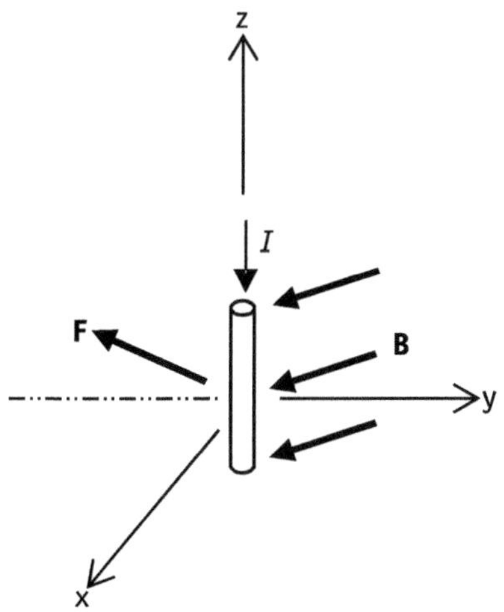

Figure 10.3 Illustration of force on a conductor carrying current.

the forces and establish equilibrium, equal and opposite forces, $\mathbf{F}_a$, would have to be applied. If motion occurs, the work done on the system by the outside agent applying the force is given by the integral:

$$W = \int_{\text{initial } l}^{\text{final } l} \mathbf{F}_a \cdot d\mathbf{l} \tag{10.12}$$

A positive result from the integration indicates that work was done by the agent on the system to move the particles or conductor from the initial location to the final location against the field. Because the magnetic force, and, hence, $\mathbf{F}_a$, is generally nonconservative, the entire path of integration joining the initial and final locations of the conductor must be specified [1].

Example 10.5

Find the work and power required to move the conductor shown in Figure 10.4 one full revolution in the direction shown after 0.02s, if $\mathbf{B} = 2.5 \times 10^{-3}\ \mathbf{a}_r$ T and the current is 45.0A [3].

Solution: See the expressions shown immediately below

$$\begin{aligned}\mathbf{F} &= I(\mathbf{L} \times \mathbf{B}) = I\left[l(-\mathbf{a}_z) \times B(\mathbf{a}_r)\right] \\ &= -1.13 \times 10^{-2}\mathbf{a}_\phi \text{ N}\end{aligned}$$

where

$I = 45.0\text{A}$

$\mathbf{L} = l(-\mathbf{a}_z) = 0.10\text{m}\ (-\mathbf{a}_z)$

$l = 0.10\text{m}$

$\mathbf{B} = B(\mathbf{a}_r) = 2.5 \times 10^{-3}\ \text{T}\ (\mathbf{a}_r)$

$B = 2.5 \times 10^{-3}\ \text{T}$

and so ...

$$\mathbf{F}_a = -1.13 \times 10^{-2}\,\mathbf{a}_\phi\ \text{N}$$

$$W = \int \mathbf{F}_a \cdot d\mathbf{l} = \int_{\phi_{\text{initial}}}^{\phi_{\text{final}}} \mathbf{F}_a \cdot d\mathbf{l}$$

$$= \int_0^{2\pi} \left(-1.13 \times 10^{-2}\right)\mathbf{a}_\phi \cdot \left(r\,d\phi\right)\mathbf{a}_\phi = -2.13 \times 10^{-3}\ \text{J}$$

where

$I = 45.0\text{A}$

$\phi_{\text{final}} = 2\pi \text{ radians} = 2\pi$

$\mathbf{L} = l(-\mathbf{a}_z) = 0.10\text{m}\ (-\mathbf{a}_z)$

$\phi_{\text{initial}} = 0 \text{ radians} = 0$

$l = 0.10\text{m}$

$\mathbf{B} = B(\mathbf{a}_r) = 2.5 \times 10^{-3}\ \text{T}\ (\mathbf{a}_r)$

$B = 2.5 \times 10^{-3}\ \text{T}$

and

$$P = \frac{W}{t} = \frac{-2.13 \times 10^{-3}\ \text{J}}{0.02\ \text{s}} = -0.107\text{W}$$

where

$W = -2.13 \times 10^{-3}\ \text{J}$

$t = 0.02\text{s}$

The negative sign means that work is done by the magnetic field in moving the conductor in the direction shown in Figure 10.4. For motion in the opposite direction, the reversed limits will provide the change of sign, and no attempt to place a sign on $rd\phi\ \mathbf{a}_\phi$ should be made [1].

10.5 TORQUE

The *moment of a force* or *torque* about a specified point is the cross-product of the *lever arm* about that point and the force. The lever arm, **r**, is directed from the point about which the torque is to be obtained to the point of application of the force. In Figure 10.5, the force at P has a torque about the origin given by:

$$\mathbf{T} = \mathbf{r} \times \mathbf{F} \tag{10.13}$$

where **T** has the units Nm (the units Nm/rad is another alternative unit to distinguish torque from energy).

In Figure 10.5, **T** lies along an axis (in the *xy*-plane) through the origin. If P were joined to the origin by a rigid rod freely pivoted at the origin, then the applied force would tend to rotate about that axis. The torque **T** would then be said to be *about the axis* rather than *about point O* [1].

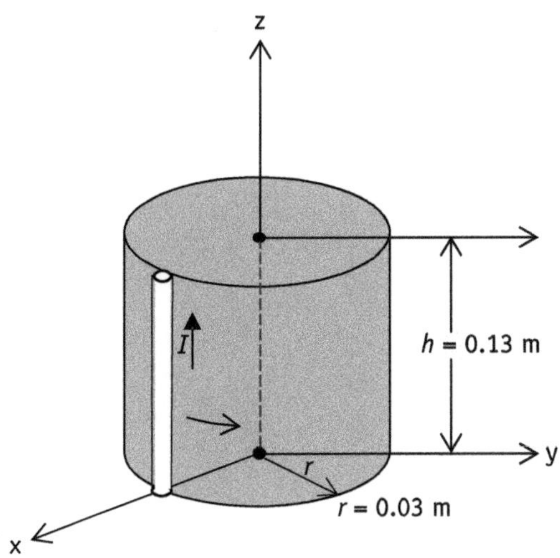

Figure 10.4 Illustration for Example 10.5 to show work and power required.

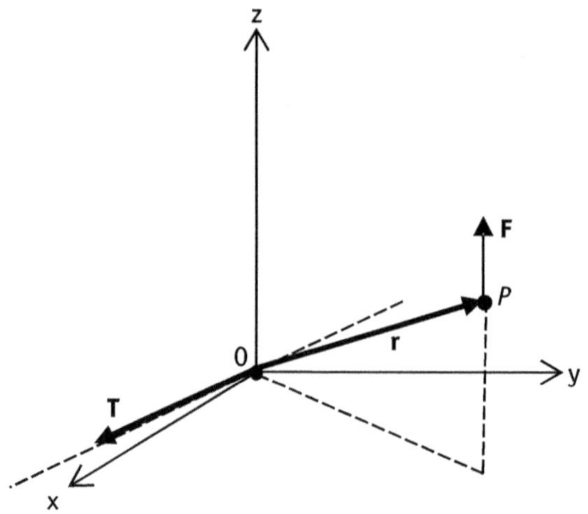

Figure 10.5 Illustration showing force at P having a torque **T** about the origin.

10.6 MAGNETIC MOMENT OF A PLANAR COIL

Consider the single-turn coil in the $z = 0$ plane shown in Figure 10.6 of width, w, in the x-direction and length, l, along y. The field **B** is uniform and in the positive x direction. The only forces result from the coil sides, l. For the side on the left,

$$\mathbf{F} = I\left(l\mathbf{a}_y \times B\mathbf{a}_x\right) = -BIl\,\mathbf{a}_x \tag{10.14}$$

and for the side on the right,

$$\mathbf{F} = BIl\,\mathbf{a}_x \tag{10.15}$$

The torque about the y-axis from the left current element requires a lever arm,

$$\mathbf{r} = -\left(\frac{w}{2}\right)\mathbf{a}_x \tag{10.16}$$

The sign will change for the lever arm to the right current element. The torque from both elements is:

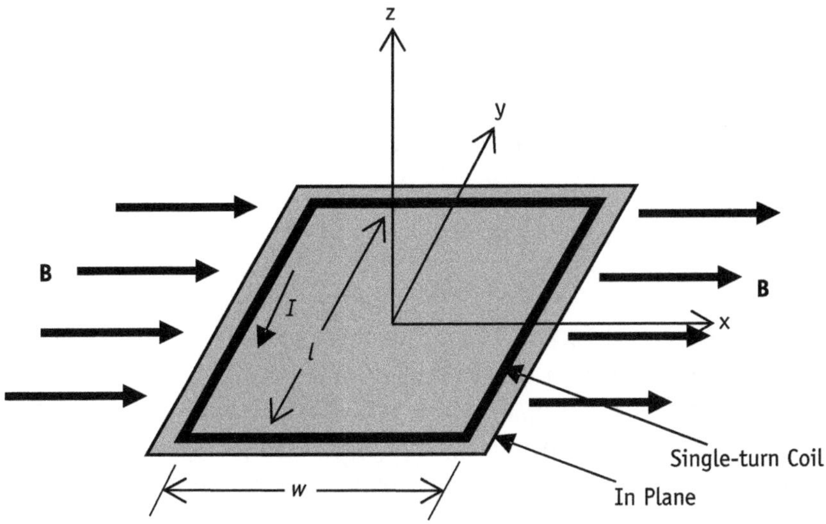

Figure 10.6 Illustration showing single-turn planar coil in the $z = 0$ plane.

$$\begin{aligned}\mathbf{T} &= \left(-\frac{w}{2}\right)\mathbf{a}_x \times (-BI\,l)\mathbf{a}_x + \left(\frac{w}{2}\right)\mathbf{a}_x \times (BI\,l)\mathbf{a}_x \\ &= BI\,l\,w\left(-\mathbf{a}_y\right) = BI\,A\left(-\mathbf{a}_y\right)\end{aligned} \tag{10.17}$$

where A is the area of the coil. It can be shown that this expression for the torque holds for a coil of arbitrary shape (and for any axis parallel to the y-axis).

The *magnetic momentum* $\mathbf{m}$ of a planar current loop is defined as $I\,A\,\mathbf{a}_n$, where the unit normal $\mathbf{a}_n$ is determined by the right-hand rule. (Note: The right thumb gives the direction of $\mathbf{a}_n$ when the fingers point in the direction of the current.) It is seen that the torque on a planar coil is related to the applied field by:

$$\mathbf{T} = \mathbf{m} \times \mathbf{B} \tag{10.18}$$

This concept of magnetic moment is essential to an understanding of the behavior of orbiting charged particles. For example, a positive charge Q moving in a circular orbit at a velocity, U, or an angular velocity, ω, is equivalent to a current

$$I = \left(\frac{\omega}{2\pi}\right)Q \tag{10.19}$$

and so gives rise to a magnetic moment

$$\mathbf{m} = \left(\frac{\omega}{2\pi}\right)QA\mathbf{a}_n \tag{10.20}$$

as shown in Figure 10.7.

More important to the discussion is the fact that, in the presence of a magnetic field, **B**, there will be a torque **T** = **m** × **B**. This torque tends to turn the current loop until **m** and **B** are in the same direction. In that orientation, the torque will be zero [4].

10.7 PROBLEMS AND SOLUTIONS

Problem 10.1

A conductor 4m long lies along the *y*-axis with a 10.0A current running in the positive *y* direction. Find the force on the conductor if the field in the region is **B** = 0.05**x** Tesla.

Problem 10.2

A current strip 2-cm wide carries a current of 15.0A in the *x*-direction as shown in Figure 10.8.

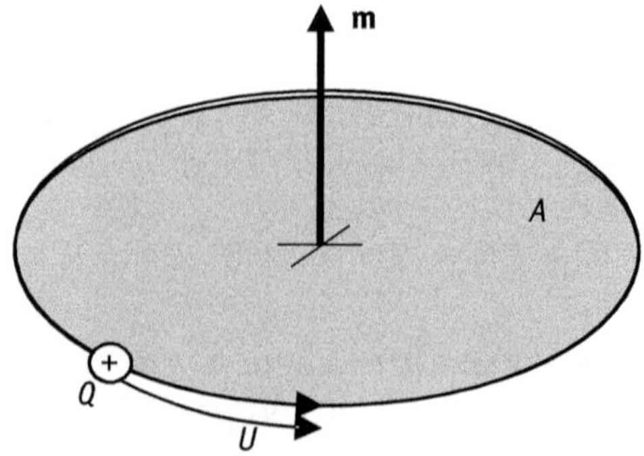

Figure 10.7 Illustration showing positive charge moving about in a circular orbit at a velocity, *U*, or an angular velocity, ω.

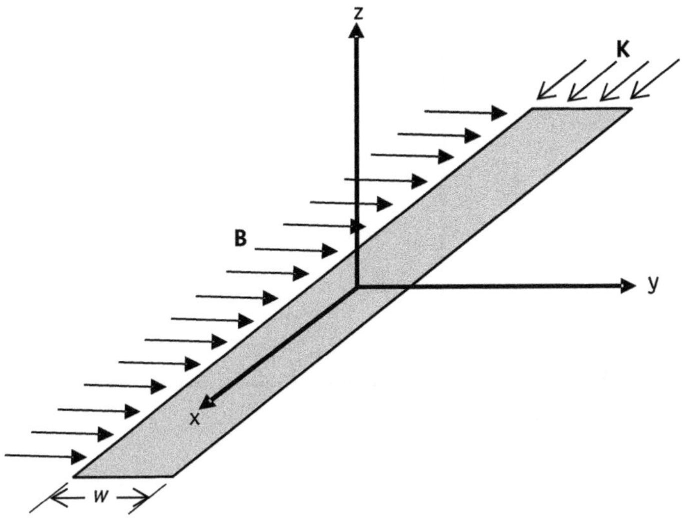

Figure 10.8 Illustration for Problem 10.2.

Find the force on the strip per unit length if the uniform B-field is $\mathbf{B} = 0.20\mathbf{y}$ T.

Problem 10.3

Find the forces per unit length on two long, straight, parallel conductors if each wire carries a current of 10.0A in the same direction. The separation distance between the wires is 0.2m. See Figure 10.9.

Problem 10.4

A conductor carries current, I, parallel to a conductive strip of current density, K_o, and width, w, as shown in Figure 10.10.

Find an expression for the force per unit length on the conductor. What is the result when the width, w, approaches ∞?

Problem 10.5

Find the torque about the y-axis for the two conductors of length, l, separated by a fixed distance, w, in the uniform field, $\mathbf{B}$, shown in Figure 10.11.

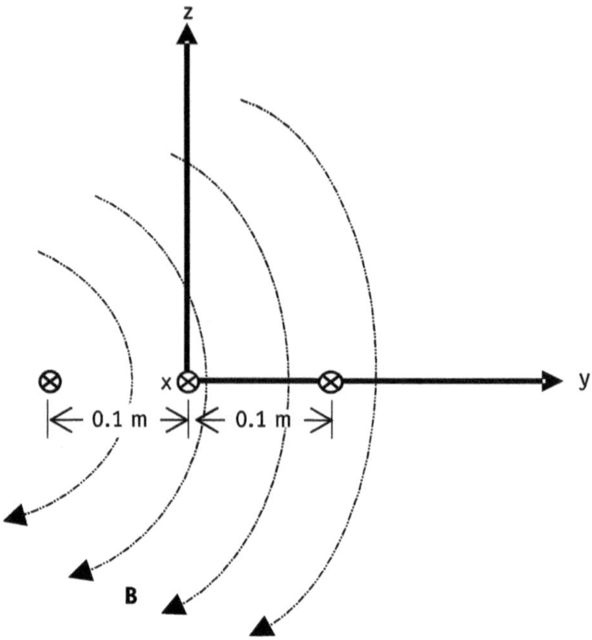

Figure 10.9 Illustration for Problem 10.3.

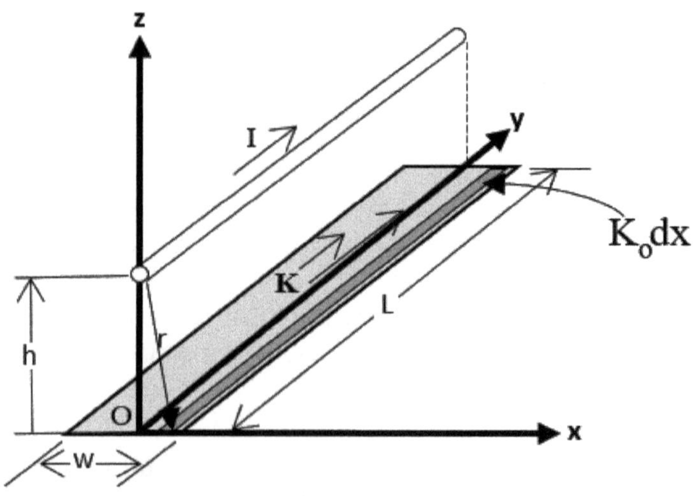

Figure 10.10 Illustration for Problem 10.4.

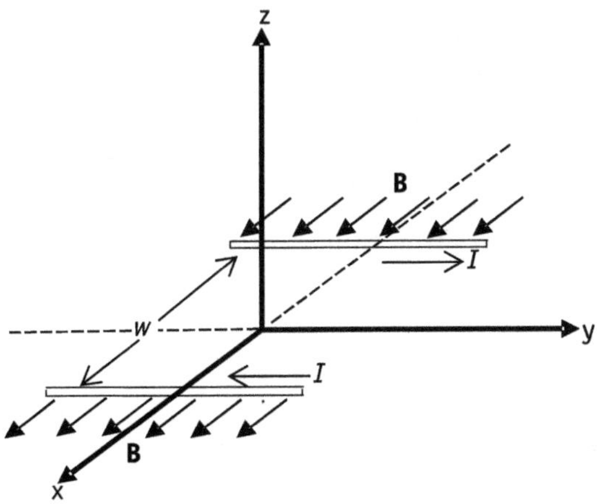

Figure 10.11 Illustration for Problem 10.5.

Problem 10.6

The rectangular coil in Figure 10.12 is in the field quantified as:

$$\mathbf{B} = 0.05\ (1/2^{1/2})\ (\mathbf{x} + \mathbf{y})\ \text{Tesla}$$

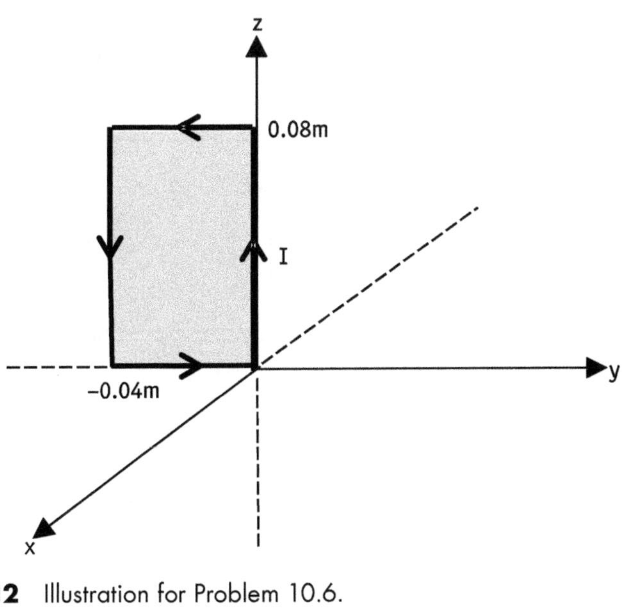

Figure 10.12 Illustration for Problem 10.6.

Find the torque about the z-axis when the coil is in the position, as shown in the figure, and carries a current of 5.0A.

Problem 10.7

Find the maximum torque on an 85-turn, rectangular coil measuring 0.2m by 0.3m and carrying a current of 2.0A in a field quantified to be B = 6.5 Tesla.

Problem 10.8

A conductor 4m long carrying a current of 10A in the $+y$ direction lies along the y-axis between the $y = \pm 2$m. If the field $\mathbf{B} = 0.05\mathbf{x}$ Tesla, find the work done moving the conductor parallel to itself at a constant speed to the position $x = z = 2.0$m. See Figure 10.13.

Problem 10.9

Find the work and power required to move the conductor shown in Figure 10.14 below one full turn in the positive direction at rotational frequency of N revolutions per minute if $\mathbf{B} = B_o\ \mathbf{r}$ Tesla (note that B_o is a positive constant).

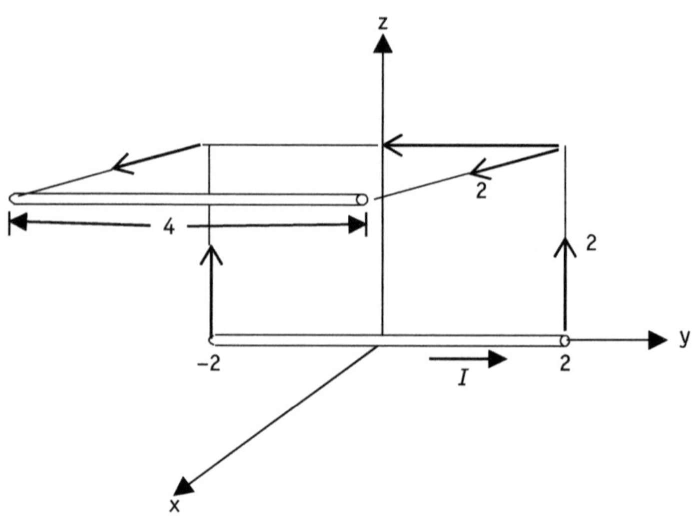

Figure 10.13 Illustration for Problem 10.8.

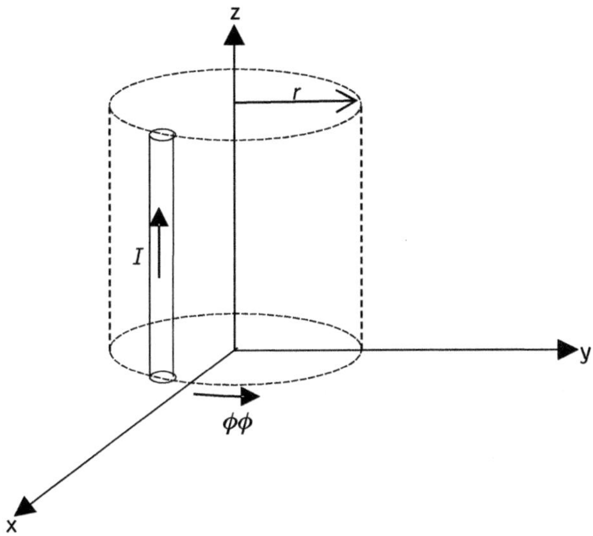

Figure 10.14 Illustration for Problem 10.9.

References

[1] Shadowitz, A., *The Electromagnetic Field*, New York: Dover Publications, 1975.

[2] Russer, P., *Electromagnetics, Microwave Circuit and Antenna Design for Communications Engineering*, Norwood, MA: Artech House, 2006.

[3] Warnick, K. F., and P. Russer, *Problem Solving in Electromagnetics, Microwave Circuit and Antenna Design for Communications Engineering*, Norwood, MA: Artech House, 2006.

[4] Cheng, D. K., *Fundamentals of Engineering Electromagnetics*, Reading, MA: Addison-Wesley, 1993.

11

INDUCTANCE AND MAGNETIC CIRCUITS

11.1 VOLTAGE OF SELF-INDUCTION

A voltage appears at the terminals of an N-turn coil in Figure 11.1, where the flux ϕ common to the turns is changing with time. This induced voltage is given by Faraday's law:

$$V = -N\left(\frac{d\phi}{dt}\right) \tag{11.1}$$

See Chapter 12 for a discussion of the polarity. Defining the *self-inductance* of the coil by:

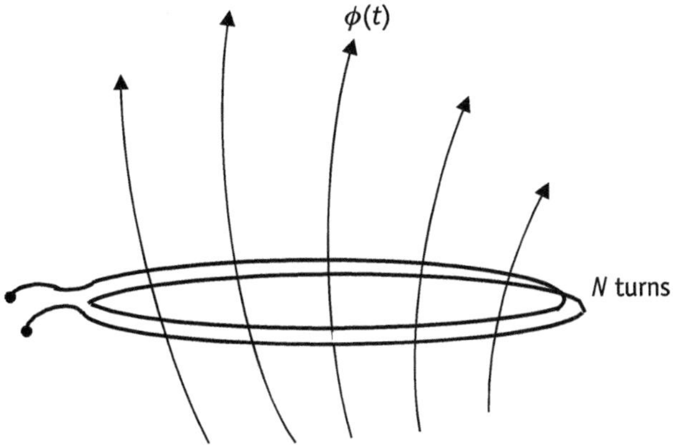

Figure 11.1 *N*-turn coil with flux ϕ.

$$L = N\left(\frac{d\phi}{di}\right) \tag{11.2}$$

Faraday's law may be written in the form:

$$V = -L\left(\frac{di}{dt}\right) \tag{11.3}$$

The units on L are henries where:

$$1\ \text{H} = 1\ \text{Wb/A} \tag{11.4}$$

In any calculations of self-inductance, account must be taken of the ferromagnetic material presence [1]. The flux, and, hence, the self-inductance, when free space is assumed will differ greatly from that which results when a portion of the region contains ferromagnetic materials.

11.2 INDUCTORS AND INDUCTANCE

An inductor (also called an inductance) is formed by two conductors separated by free space, where the conductor arrangement is such that magnetic flux from one links with the other. For static (or, at most, low-frequency) current I in the conductors, let the total flux linking the conductors be:

$$\lambda = -\left|\begin{array}{ll} N\Phi & \text{for coils} \\ \Phi & \text{for other arrangements} \end{array}\right. \tag{11.5}$$

Then, the inductance of the inductor is defined to be:

$$L = \frac{\lambda}{I} \tag{11.6}$$

It should be noted that L will always be the product of μ_0 and a geometrical factor having the dimensions of length. Compare with expressions for resistance, R (Chapter 6) and capacitance, C (Chapter 7) [1].

The expressions for the stored energy in the magnetic field from circuit analysis and field theory provide another defining expression for inductance, L. The equations [2],

$$W = \frac{1}{2} L I^2 \tag{11.7}$$

and [3],

$$W = \frac{1}{2} \int_V (\mathbf{B} \cdot \mathbf{H})\, dv \tag{11.8}$$

give:

$$L = \int_V (\mathbf{B} \cdot \mathbf{H})\, dv / I^2 \tag{11.9}$$

11.3 STANDARD FORMS

In addition to coaxial conductors, several of the more common conductor configurations for which the inductance is required are shown in Figures 11.2 through 11.8 [2].

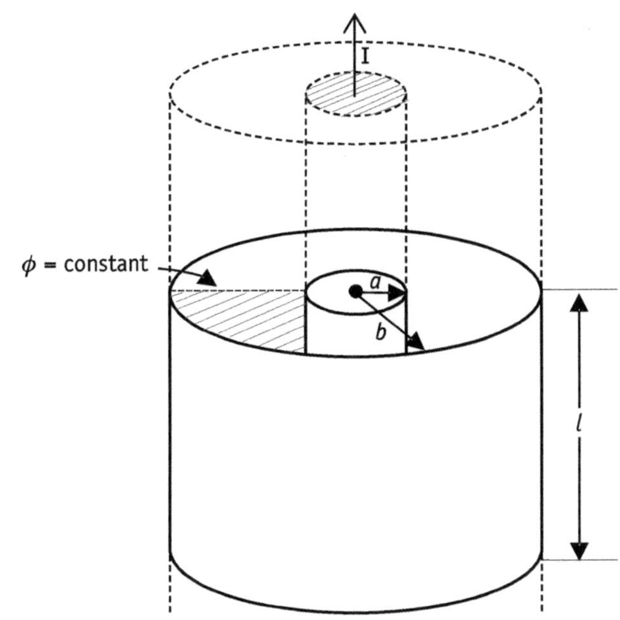

Figure 11.2 Illustration of coaxial conductors.

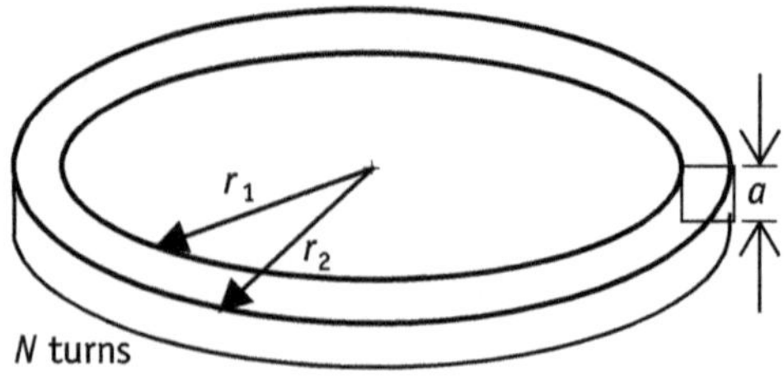

Figure 11.3 Illustration of toroid, square cross-section.

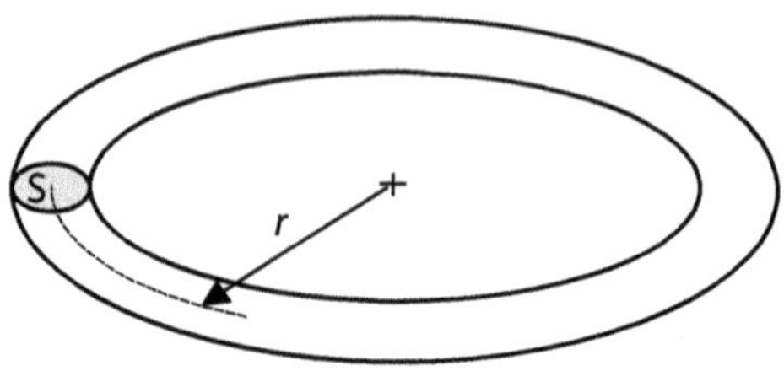

Figure 11.4 Illustration of toroid, general cross section *S*.

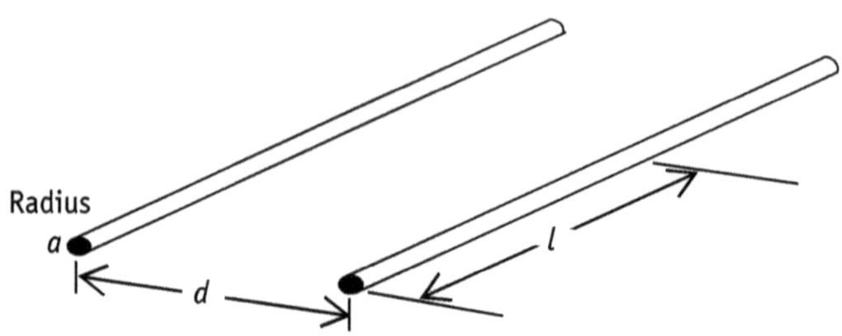

Figure 11.5 Illustration of parallel conductors of radius *a*.

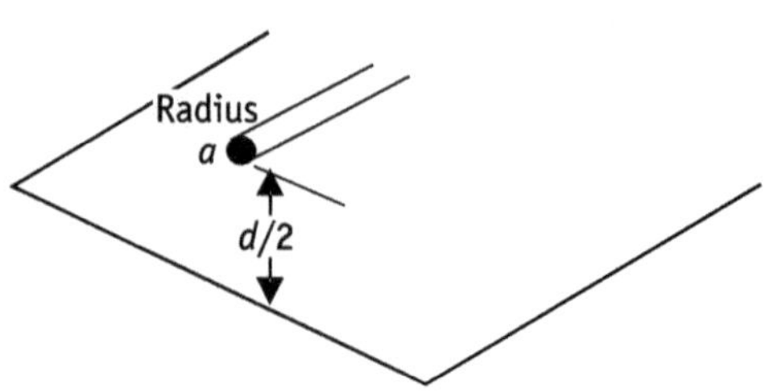

Figure 11.6 Illustration of cylindrical conductor parallel to a ground.

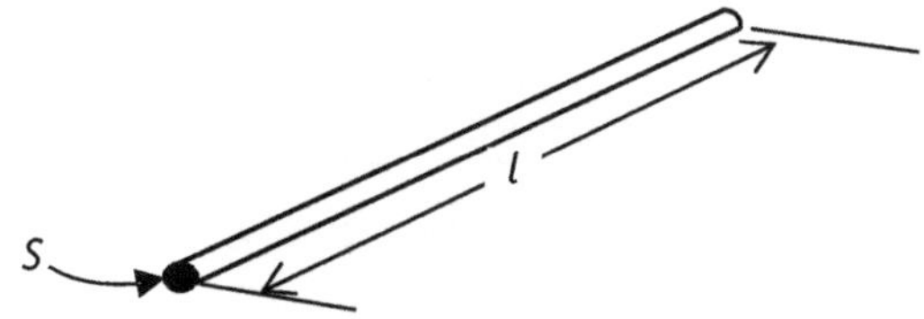

Figure 11.7 Long solenoid of small cross-sectional area, *S*.

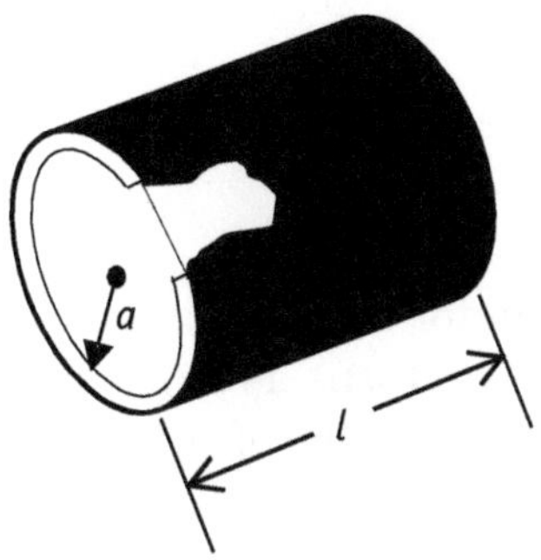

Figure 11.8 Single layer, air-core coil.

11.4 INTERNAL INDUCTANCE

Magnetic flux occurs within a conductor cross section as well as external to the conductor. This internal flux gives rise to an *internal inductance* which is often small compared to the external inductance and frequency ignored. In Figure 11.9 and Figure 11.10(a, b) a conductor of circular cross section is shown, with a current I assumed to be uniformly distributed over the area. (Note: this assumption is only valid at low frequencies seeing how *skin effect* at higher frequencies forces the current to be concentrated at the outer surface.) Within the conductor of radius, a, Ampere's law gives

$$\mathbf{H} = \left(\frac{Ir}{2\pi a^2}\right)\mathbf{a}_\phi \tag{11.10}$$

and,

$$\mathbf{B} = \mu_0\left(\frac{Ir}{2\pi a^2}\right)\mathbf{a}_\phi \tag{11.11}$$

Figure 11.9 Multilayer, air-core coil.

The piece of conductor in Figure 11.10(a) must be imagined as a short section of an infinite torus as suggested in Figure 11.10(b). The current filaments become circles of infinite radius. The lines of flux $d\Phi$ through the strip $l\ dr$ encircle only those filaments whose distance from the conductor axis is smaller than r. Thus, an open surface bounded by one of those filaments is cut once (or an odd number of times) by the lines of $d\Phi$; whereas for a filament such as *1* or *2*, the surface is cut zero times (or an even number of times). It follows that $d\Phi$ links only with the fraction $\pi r^2/\pi a^2$ of the total current so that the total flux linkage is given by the weighted sum,

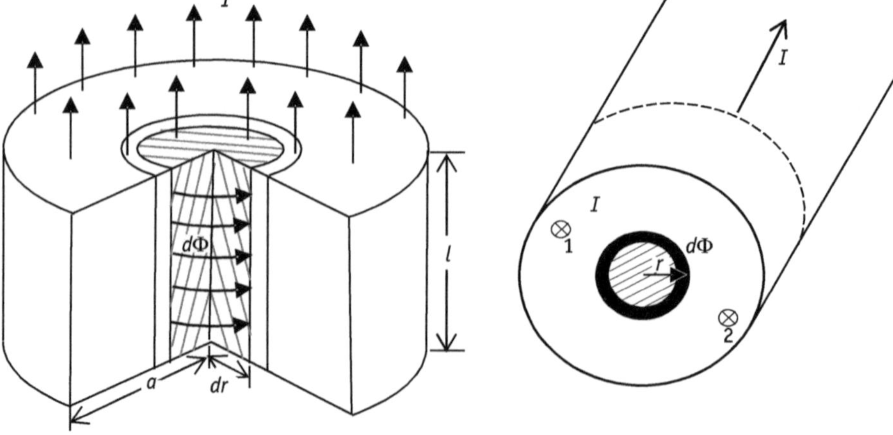

Figure 11.10 Conductor of circular coaxial cross section.

$$\lambda = \int \left(\frac{\pi r^2}{\pi a^2} \right) d\Phi = \int_0^a \left(\frac{\pi r^2}{\pi a^2} \right) \left(\frac{\mu_0 I r}{2\pi a^2} \right) l \, dr = \left(\frac{\mu_0 I l}{8\pi} \right) \tag{11.12}$$

and

$$\frac{L}{l} = \frac{(\lambda I)}{l} = \frac{\mu_0}{8\pi} = \frac{1}{2} \times 10^{-7} \text{ H/m} \tag{11.13}$$

The result is independent of the conductor radius. The total inductance is the sum of the external and internal inductances. If the external inductance is the of the order $1/2 \times 10^{-7}$ H/m, the internal inductance should not be ignored [3].

11.5 MAGNETIC CIRCUITS

In Chapter 9, magnetic field intensity **H**, flux **Φ**, and magnetic flux density **B** were examined where the medium was free space. For example, when Ampere's law is applied to the closed path C through the long, air-core shown in Figure 11.11.

The result is:

$$\oint \mathbf{H} \cdot d\mathbf{l} = NI \tag{11.14}$$

But since the flux lines are widely spread outside of the coil, B is small there. The flux is effectively restricted to the inside of the coil where:

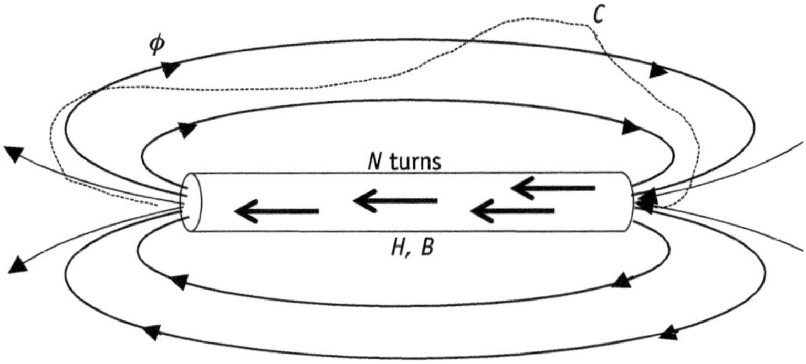

Figure 11.11 Ampere's law applied to the closed path, C.

$$H \cong \frac{NI}{l} \tag{11.15}$$

Ferromagnetic materials have relative permeabilities, μ_r, in the order of thousands.

Consequently, the flux density:

$$B = \mu_0 \mu_r H \tag{11.16}$$

is for a given H much greater than would result in free space. In Figure 11.12, the coil is not distributed over the entire iron core. Even so, the NI of the coil causes a flux Φ which follows the core. It might be said that the flux prefers the core to the surrounding space by the ratio of several thousand to one.

This is so different from the free-space magnetics of Chapter 9 that an entire subject area, known as iron core magnetics or magnetic circuits, was developed. This brief introduction to the subject assumes that all of the flux is within the core. It is also assumed that the flux is uniformly distributed over the core's cross section. Core lengths required for calculation of NI drops are mean lengths [4].

11.6 NONLINEARITY OF THE B-H CURVE

A sample of ferromagnetic material could be tested by applying increasing values of H and measuring the corresponding values of flux density B. *Magnetization curves*, or simply, *B-H curves*, for some common ferromagnetic materials are given in Figures 11.13 and 11.14.

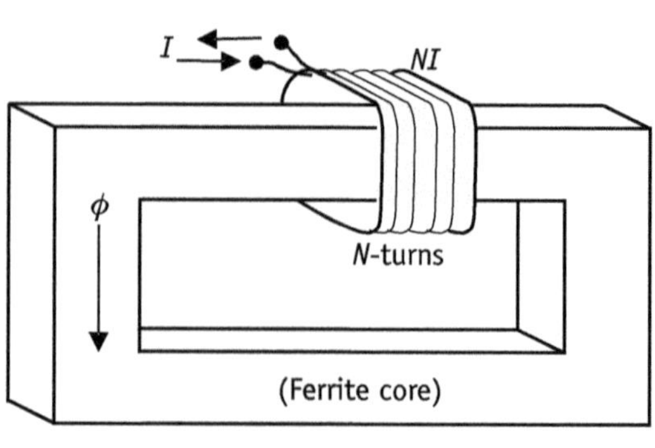

Figure 11.12 Electric conductors wrapped around a ferromagnetic core.

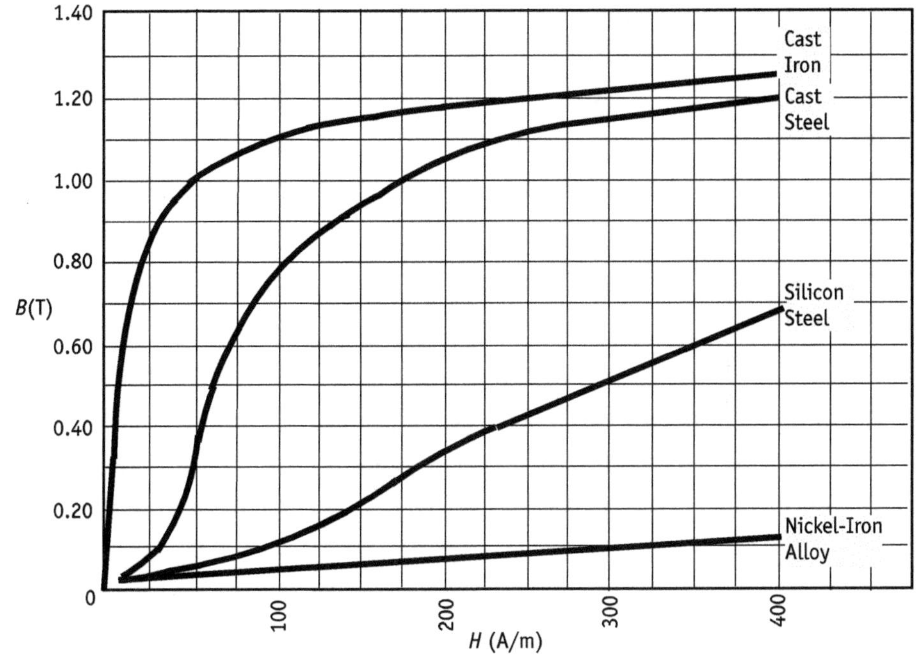

Figure 11.13 Magnetization (B-H) curves for four materials. $H < 400$ A/M.

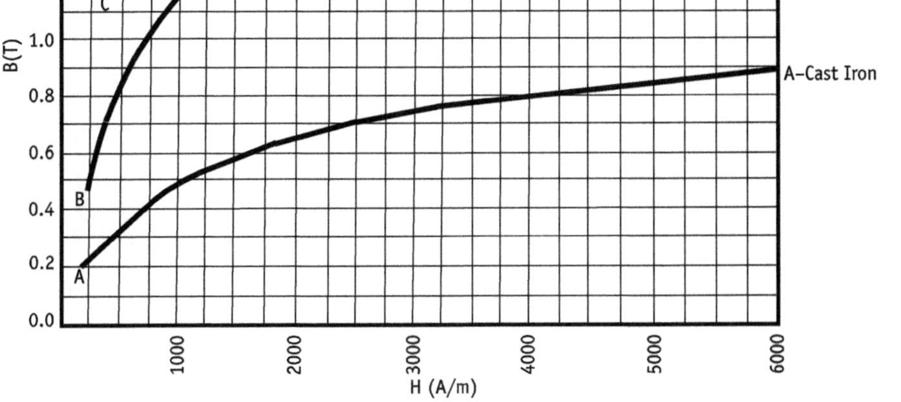

Figure 11.14 Magnetization (B-H) curves for four materials. $H > 400$ A/M.

The relative permeability can be computed from the *B-H* curve by use of both the curve and the equation

$$\mu_r = \frac{B}{(\mu_0 H)} \tag{11.17}$$

Figure 11.15 shows the extreme nonlinearity of μ_r versus H for silicon steel. This nonlinearity requires that problems be solved graphically [1].

11.7 AMPERE'S LAW FOR MAGNETIC CIRCUITS

A coil of N turns and current I around a ferromagnetic core produces a *magnetomotive force* (mmf) given by NI. The symbol F is sometimes

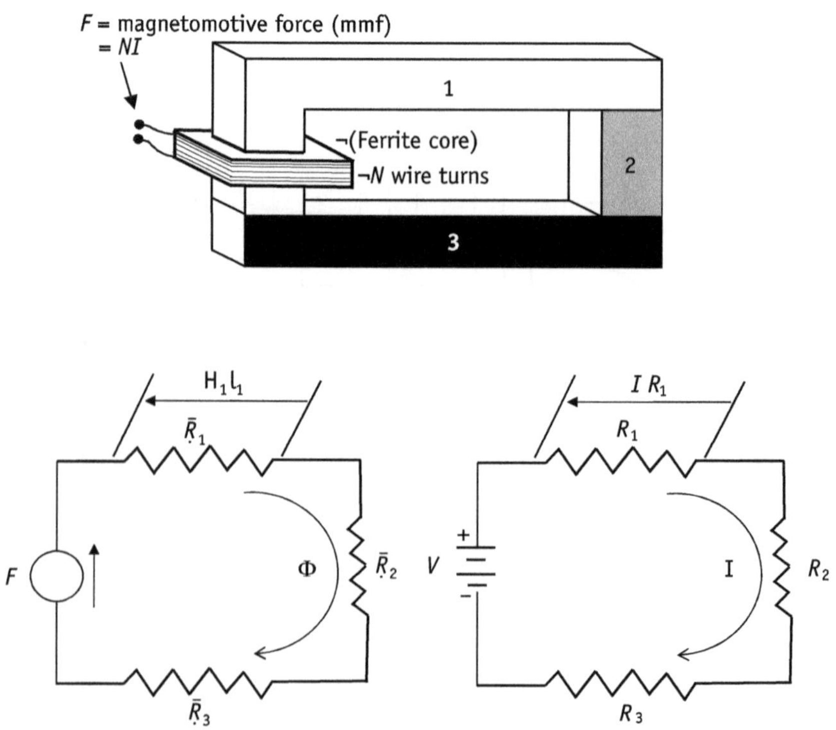

Figure 11.15 Ampere's law for magnetic circuits.

used for this mmf. The units are amperes (A) Ampere's law applied around the path in the center of the core shown in Figure 11.15(a), gives:

$$\begin{aligned} F = NI = \oint \mathbf{H} \cdot d\mathbf{l} = \int_1 \mathbf{H} \cdot d\mathbf{l} = \int_2 \mathbf{H} \cdot d\mathbf{l} = \int_3 \mathbf{H} \cdot d\mathbf{l} \\ = H_1 l_1 + H_2 l_2 + H_3 l_3 \end{aligned} \tag{11.18}$$

Comparison with Kirchhoff's law around a single closed loop with three resistors and an emf *V*,

$$V = V_1 + V_2 + V_3 \tag{11.19}$$

suggests that *F* can be viewed as an *NI* rise and the *Hl* terms considered *NI* drops in analogy to the voltage rise *V* and voltage drops V_1, V_2, and V_3. The analogy is developed in Figures 11.15(b) and 11.15(c). Flux **Φ** in Figure 11.15(b) is analogous to current *I* (see Figure 11.15(c)).

Reluctance R is analogous to electrical resistance *R*. An expression for reluctance can be developed as follows:

$$NI \text{ drop} = Hl = BA\left(\frac{l}{\mu A}\right) = \Phi \bar{R} \tag{11.20}$$

Hence,

$$\bar{\mathbf{R}} = \left[\frac{l}{(\mu A)}\right] \left(H^{-1}\right) \tag{11.21}$$

If the reluctances are known, then the equation

$$F = NI = \Phi\left(\bar{R}_1 + \bar{R}_2 + \bar{R}_3\right) \tag{11.22}$$

can be written for the magnetic circuit of Figure 11.15(b). However, μ_r must be known for each material before reluctance can be calculated, and only after *B* or *H* is known will the value of μ_r be known. This is in contrast to the relation

$$R = \frac{l}{\sigma A} \tag{11.23}$$

See Section 6.7 in which the conductivity, σ, was independent of the current.

11.8 CORES WITH AIR GAPS

Magnetic circuits with small air gaps are very common. The gaps are generally kept as small as possible since the *NI* drop of the air gap is often much greater than the drop in the core. The flux fringes outward at the gap so that the area at the gap exceeds the area of the adjacent core. Provided that the gap length, l_a, is less than 1/10 the smaller dimension of the core, an *apparent area*, S_a, of the air gap can be calculated. For a rectangular core of dimensions a and b,

$$S_a = (a + l_a)(b + l_a) \tag{11.24}$$

If the total flux in the air gap is known, H_a and $H_a l_a$ can be computed directly.

$$H_a = \frac{1}{\mu_0}\left(\frac{\Phi}{S_a}\right) \tag{11.25}$$

$$H_a l_a = l_a\left(\frac{1}{\mu_0}\right)\left(\frac{\Phi}{S_a}\right) = \left(\frac{l_a \Phi}{\mu_0 S_a}\right) \tag{11.26}$$

For a uniform iron core of length, l_i, with a single air gap, Ampere's law reads:

$$NI = H_i l_i + H_a l_a = H_i l_i + \frac{l_a \Phi}{\mu_0 S_a} \tag{11.27}$$

If the flux Φ is known, it is not difficult to compute the *NI* drop across the air gap, obtain B_i, take H_i from the appropriate *B-H* curve and compute the *NI* drop in the core, $H_i l_1$. The sum is the *NI* required to establish the flux Φ. However, with *NI* given, it is a matter of trial and error to obtain B_i and Φ.

11.9 MULTIPLE COILS

Two or more coils on a core could be wound such that their mmfs either aid one another or oppose. Consequently, a method of indicating polarity is given in Figure 11.16.

An assumed direction for the resulting flux Φ could be incorrect just as an assumed current in a dc circuit with two or more voltage sources may be incorrect. A negative result simply means that the flux is in the opposite direction [3].

11.10 PARALLEL MAGNETIC CIRCUITS

The method of solving a parallel magnetic circuit is suggested by the two-loop equivalent circuit shown in Figure 11.17(a, b).

The leg on the left contains an *NI* rise and an *NI* drop. The *NI* drop between the junctions a and b can be written for each leg as follows:

$$F - H_1 l_1 = H_2 l_2 = H_3 l_3 \tag{11.28}$$

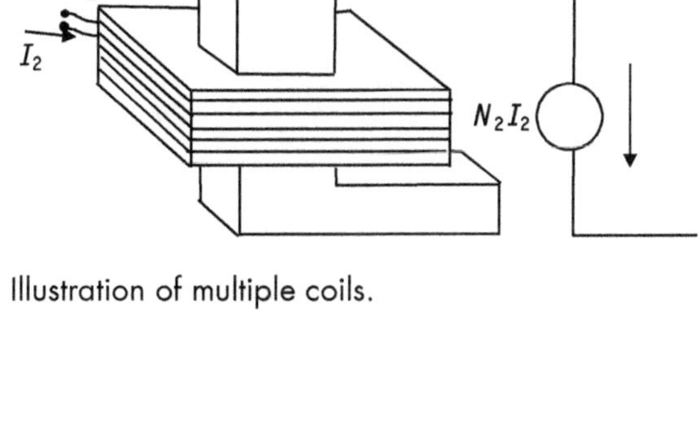

Figure 11.16 Illustration of multiple coils.

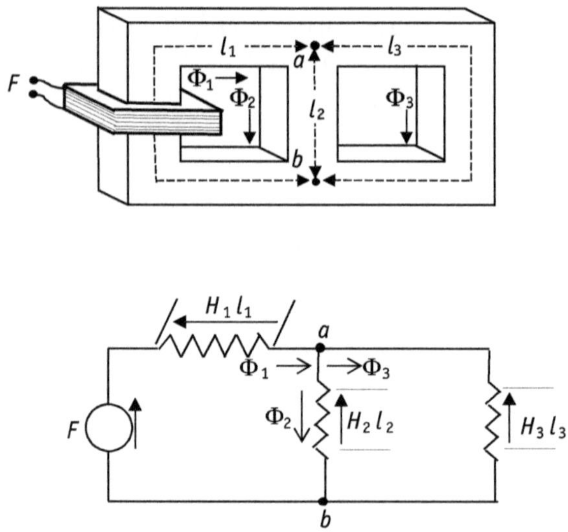

Figure 11.17 Illustration of parallel magnetic circuits.

and the fluxes satisfy

$$\Phi_1 = \Phi_2 + \Phi_3 \tag{11.29}$$

Different materials for the core parts will necessitate working with several B-H curves. An air gap in one of the legs would lead to $H_i l_i + H_a l_a$ for the mmf between the junctions for that leg.

The equivalent magnetic circuit should be drawn for parallel magnetic circuit problems. It is good practice to mark the material types, cross-sectional areas, and mean length directly on the diagram. In more complex problems, a scheme like Table 11.1 can be helpful. The data are inserted directly into the table. The remaining

Table 11.1
Data Collection Matrix for Magnetic Circuit Analysis

Part	Material	Area	l	Φ	B	H	Hl
#1							
#2							
#3							

quantities are then calculated or taken from the appropriate *B-H* curve [4].

11.11 PROBLEMS AND SOLUTIONS

Problem 11.1

Find the inductance per unit length of the coaxial transmission line given in Figure 11.18.

Problem 11.2

Find the inductance per unit length of the parallel cylindrical conductors (see Figure 11.19) where $l = 25$ ft and radius $a = 0.803$ in.

Problem 11.3

A circular conductor with the same radius, a, as in the previous problem, is 12.5 feet from an infinite conducting plane.

Find the inductance (see Figure 11.20).

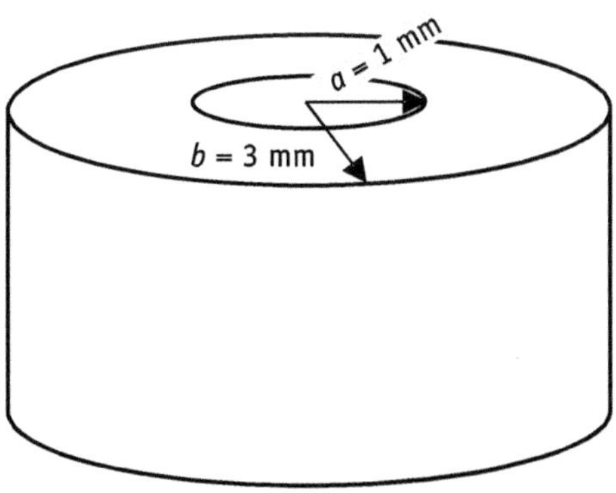

Figure 11.18 Illustration for Problem 11.1.

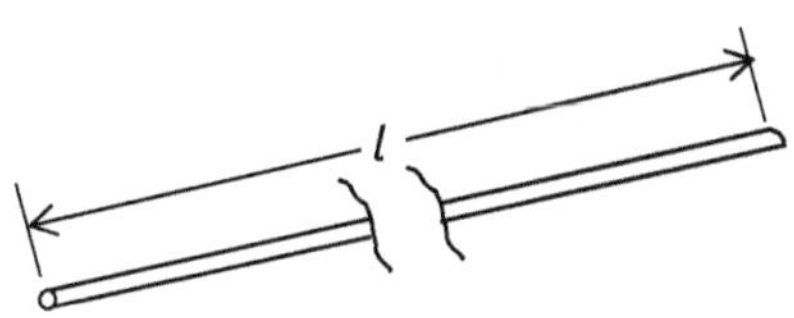

Figure 11.19 Illustration for Problem 11.2.

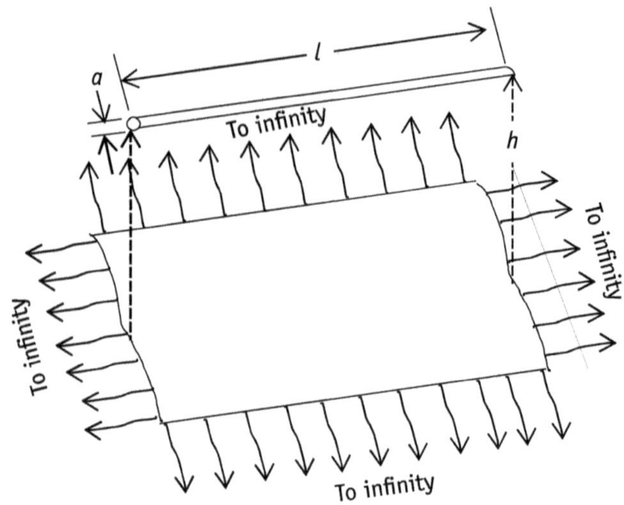

Figure 11.20 Illustration for Problem 11.3.

Problem 11.4

An air-core solenoid of 300 turns and length 0.5m has a single layer of conductor at a radius 0.02m. Find the inductance.

Problem 11.5

Find the inductance of the coil shown in Figure 11.21 where $N = 300$, $r_1 = 9$ mm, $r_2 = 25$ mm, and $l = 20$ mm.

Problem 11.6

Assume that the air-core toroid shown in Figure 11.22 has a circular cross section of radius 4 mm.

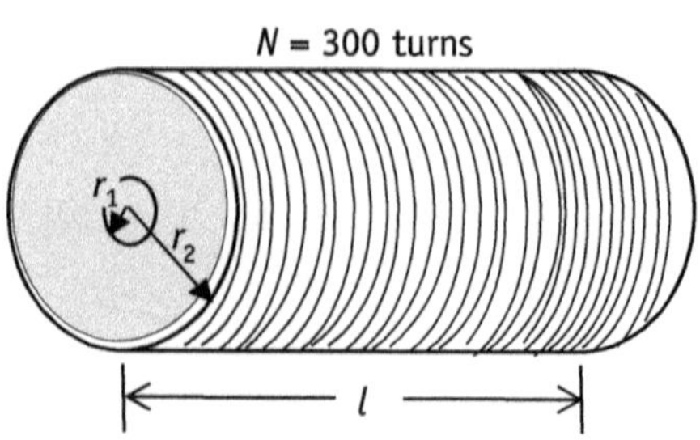

Figure 11.21 Illustration for Problem 11.5.

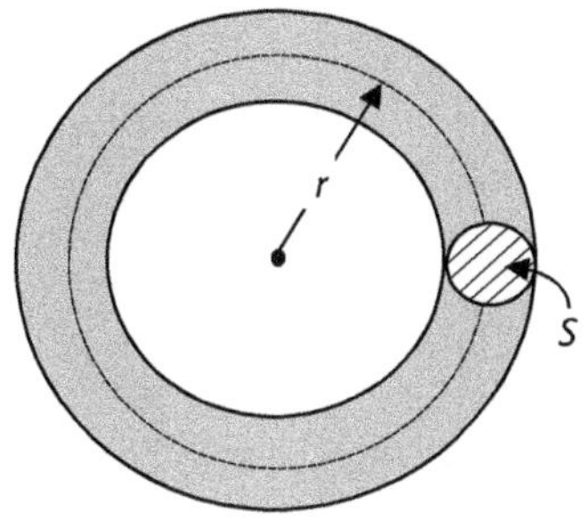

Figure 11.22 Illustration for Problem 11.6.

Find the inductance if there are 2,500 turns and the mean radius is $r = 20$ mm.

Problem 11.7

Assume the air-core toroids of Figure 11.23 have 700 turns, an inner radius of 1 cm, an outer radius of 2 cm, and a height = 1.5 cm. Find L using the appropriate equations.

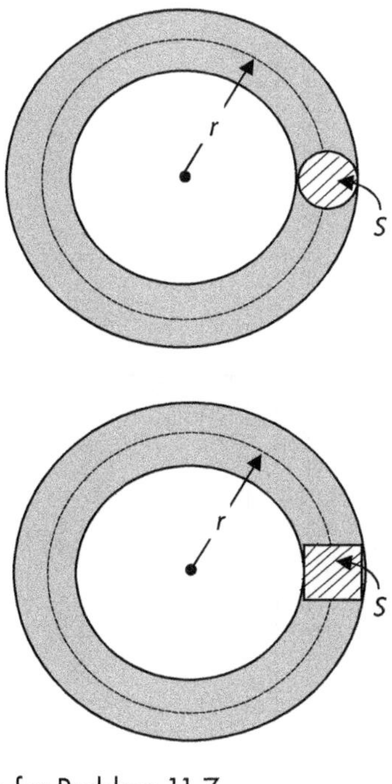

Figure 11.23 Illustration for Problem 11.7.

(a) The formula for square cross-section toroids;

(b) The approximate formula for general toroid which assumes a uniform H at a mean radius.

Problem 11.8

Use the energy integral to find the internal inductance per unit length of a length of a cylindrical conductor of radius a (see Figure 11.24).

Problem 11.9

The cast iron core shown in Figure 11.25 has an inner radius of 7 cm and an outer radius of 9 cm. Find the flux, Φ, if the coil mmf (F) is 500A.

Problem 11.10

The magnetic circuit shown in Figure 11.26 is cast iron with a mean length, l_1, of 0.44m and square cross section of 0.02m × 0.02m. The air gap length, l_a, 2 mm and the coil contains 400 turns.

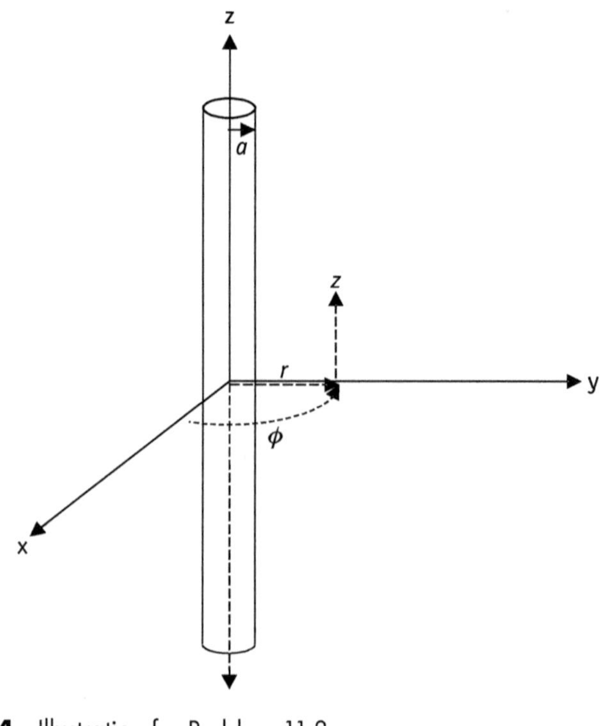

Figure 11.24 Illustration for Problem 11.8.

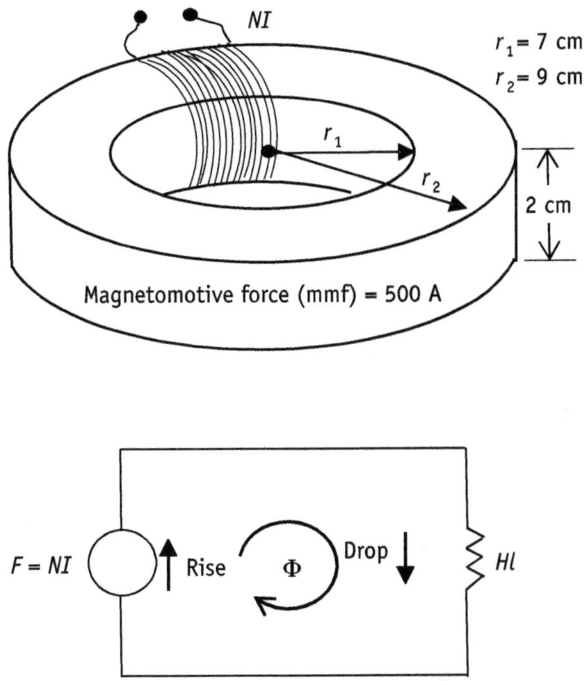

Figure 11.25 Illustration for Problem 11.9.

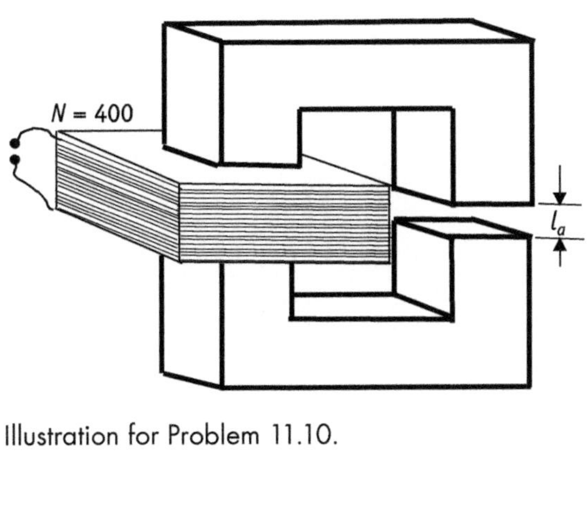

Figure 11.26 Illustration for Problem 11.10.

Find the current I required to establish an air gap flux of 0.141 mWb.

Problem 11.11

Determine the reluctance of an air gap in a dc machine in which the apparent area is:

$$S_a = 4.26 \times 10^{-2}\ \mathrm{m}^2$$

and the gap length is $l_a = 5.6$ mm.

References

[1] Shadowitz, A., *The Electromagnetic Field*, New York: Dover Publications, 1975.

[2] Russer, P., *Electromagnetics, Microwave Circuit and Antenna Design for Communications Engineering*, Norwood, MA: Artech House, 2006.

[3] Warnick, K. F., and P. Russer, *Problem Solving in Electromagnetics, Microwave Circuit and Antenna Design for Communications Engineering*, Norwood, MA: Artech House, 2006.

[4] Cheng, D. K., *Fundamentals of Engineering Electromagnetics*, Reading, MA: Addison-Wesley, 1993.

12

DISPLACEMENT CURRENT AND INDUCED EMF

12.1 DISPLACEMENT CURRENT

For static fields, the curl of **H** is equal to **J**, the current density at the point. This current density, which is due to the motion of actual charges (i.e., electrons, protons, and ions), will be called the *conduction current* density and will be denoted $\mathbf{J}_C$. It is clear that, if

$$\nabla \times \mathbf{H} = \mathbf{J}_c \tag{12.1}$$

is also held for time varying fields, then

$$\nabla \cdot \mathbf{J}_c = \nabla \cdot (\nabla \times \mathbf{H}) = 0 \tag{12.2}$$

(See Section 9.4.)

This would be incompatible with the continuity equation,

$$\nabla \cdot \mathbf{J}_c = -\frac{\partial \rho}{\partial t} \tag{12.3}$$

obtained in Section 6.9. James Clerk Maxwell postulated that [1]:

$$\nabla \times \mathbf{H} = \mathbf{J}_c + \mathbf{J}_D \tag{12.4}$$

where the *displacement current* density, $\mathbf{J}_D$, is defined by

$$\mathbf{J}_D = \frac{\partial \mathbf{D}}{\partial t} \tag{12.5}$$

With the displacement current added to the conduction current, the continuity equation is saved. Indeed,

$$\nabla \cdot \mathbf{J}_c = \nabla \cdot (\nabla \times \mathbf{H}) - \nabla \cdot \mathbf{J}_D = 0 - \nabla \cdot \frac{\partial \mathbf{D}}{\partial t} = -\frac{\partial}{\partial t}(\nabla \cdot \mathbf{D}) = -\frac{\partial \rho}{\partial t} \tag{12.6}$$

where $\nabla \cdot \mathbf{D} = \rho$ (Section 4.3) has been utilized.

The displacement current (in amperes) through a specified open surface is obtained by integration and in the same way as is the conduction current. Thus,

$$i_c = \int_S \mathbf{J}_c \cdot d\mathbf{S} \tag{12.7}$$

$$i_D = \int_S \mathbf{J}_D \cdot d\mathbf{S} = \int_S \frac{\partial \mathbf{D}}{\partial t} \cdot d\mathbf{S} \tag{12.8}$$

The expression for i_D may be interpreted in terms of the motion of charges. If charge Q is moving with velocity $\mathbf{U}$, the electric field surrounding Q also advances with velocity $\mathbf{U}$.

Therefore, even though Q may not be physically crossing a surface S (which would constitute a conduction current), the lines of $\mathbf{D}$ will be changing across S, resulting in a displacement current [1].

Example Problem

A time-varying voltage applied to a parallel plate capacitor (see Figure 12.1) [2] results in a time-varying current, i_c, in the connecting leads. Two open surfaces with a common contour C are shown [1]. Because $\nabla \cdot (\nabla \times \mathbf{H}) = 0$, the divergence theorem (Section 4.5) gives

$$\int_{S_1} (\nabla \times \mathbf{H}) \cdot d\mathbf{S} = \int_{S_z} (\nabla \times \mathbf{H}) \cdot d\mathbf{S}$$

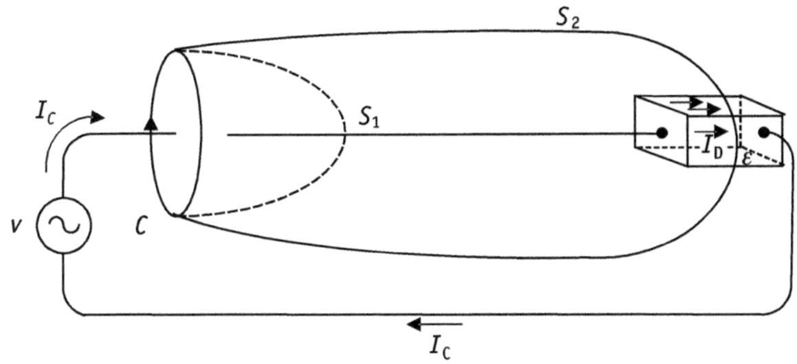

Figure 12.1 Configuration with two open surfaces having a common contour.

or

$$\int_{S_1}\left[\mathbf{J}_c + \left(\frac{\partial \mathbf{D}}{\partial t}\right)\right]\cdot d\mathbf{S} = \int_{S_2}\left[\mathbf{J}_c + \left(\frac{\partial \mathbf{D}}{\partial t}\right)\right]\cdot d\mathbf{S}$$

In the absence of fringing, **D** will be changing only within the capacitor. Moreover, there can be no moving charges ($J_C = 0$) in the dielectric. Therefore,

$$\int_{S_1}\mathbf{J}_c \cdot d\mathbf{S} = \int_{S_2}\left(\frac{\partial \mathbf{D}}{\partial t}\right)\cdot d\mathbf{S}$$

where S_3 is that part of S_2 which lies in the dielectric. The integral on the left is simply i_c, the conduction current, made up of the moving charges in the leads. The integral on the right is the displacement current, i_D, in the dielectric.

The equality of i_c and i_D for this case is demonstrated in Problem 12.1.

12.2 RATIO OF J_C TO J_D

Some materials are neither good conductors nor perfect dielectrics such that both conduction current and displacement current exist. A model for the poor conductor or lossy dielectric is shown in Figure 12.2.

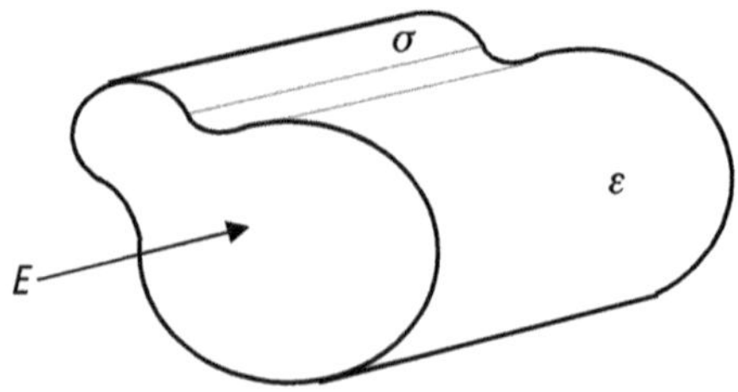

Figure 12.2 Model for the poor conductor or lossy dielectric.

Assuming the time-dependent $e^{j\omega t}$ for **E**, the total current density is:

$$\mathbf{J}_t = \mathbf{J}_c + \mathbf{J}_D = \sigma\mathbf{E} + \frac{\partial}{\partial t}(\varepsilon\mathbf{E}) = \sigma\mathbf{E} + j\omega\varepsilon\mathbf{E} \tag{12.9}$$

From which

$$\frac{\mathbf{J}_c}{\mathbf{J}_D} = \frac{\sigma}{\omega\varepsilon} \tag{12.10}$$

As expected, the displacement current becomes increasingly important as the frequency increases [2].

12.3 FARADAY'S LAW

When a conductor moves through a magnetic field cutting a flux, a voltage is induced in the conductor. Similarly, when the flux cuts across a stationary conductor, a voltage is induced. In either case, the voltage and the rate of cutting the flux are related by Faraday's law:

$$v = -\frac{d\phi}{dt} \tag{12.11}$$

Polarity of the induced voltage is sometimes illustrated by a distorted flux field as shown in Figure 12.3. The lines appear to be pushed ahead of the moving conductor.

A counterclockwise flux is shown around the conductor which would create the same distorted field. By the right-hand rule, the

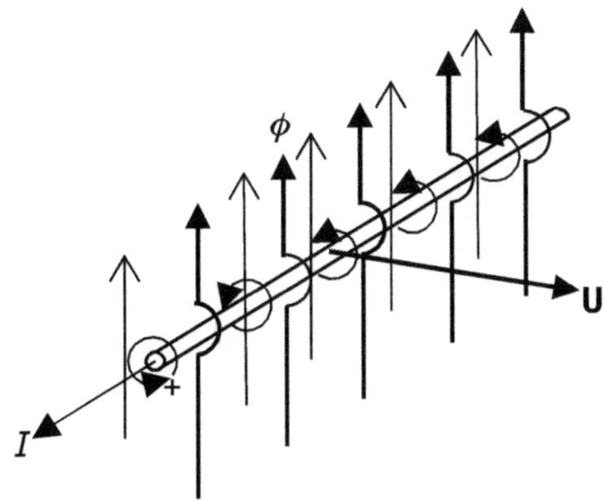

Figure 12.3 Distorted flux field modeling the polarity of the induced voltage.

current, if a closed path were provided, would leave the conductor as shown. Figure 12.4(a) shows the current direction, through an external circuit. Figure 12.4(b) shows an equivalent voltage source and the same current direction through the external circuit.

The negative sign in Faraday's law can be explained by rewriting the equation in integral form.

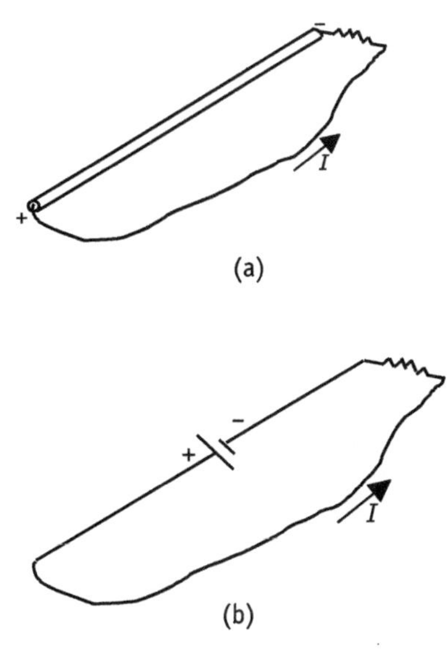

Figure 12.4 (a) Current direction through an external circuit, and (b) equivalent voltage source and same current direction through the external circuit.

$$v = -\frac{d\varphi}{dt} \tag{12.12}$$

$$\oint \mathbf{E} \cdot d\mathbf{l} = -\frac{d}{dt}\int_S \mathbf{B} \cdot d\mathbf{S} \tag{12.13}$$

The contour C around which the integral on the left is taken encloses the plane area S of the surface integral on the right. A positive direction is assigned to C and the direction of the normal, $d\mathbf{S}$, is then determined by the familiar right-hand rule. See Figure 12.5(a).

A changing magnetic flux density, $\mathbf{B}(t)$, is present within the contour. If $\mathbf{B}$ is increasing with time, the time derivative will be positive and, thus, the right side of the equation will be negative. In order for the left integral to be negative, the direction of $\mathbf{E}$ must be opposite to that of the contour as in Figure 12.5(b). A conducting filament in place of the contour would carry a current i_c also in the direction of $\mathbf{E}$.

As shown in Figure 12.5(c), such a current loop generates a flux ϕ' which opposes the increase in $\mathbf{B}$.

Lenz's law summarizes this involved discussion: The voltage induced by a changing flux has a polarity such that the current established in a closed path gives rise to a flux that opposes the change in flux.

In the special case of a conductor moving through a stationary magnetic field, the polarity predicted by Lenz's law will always be such that the conductor experiences magnetic forces that oppose its motion [3].

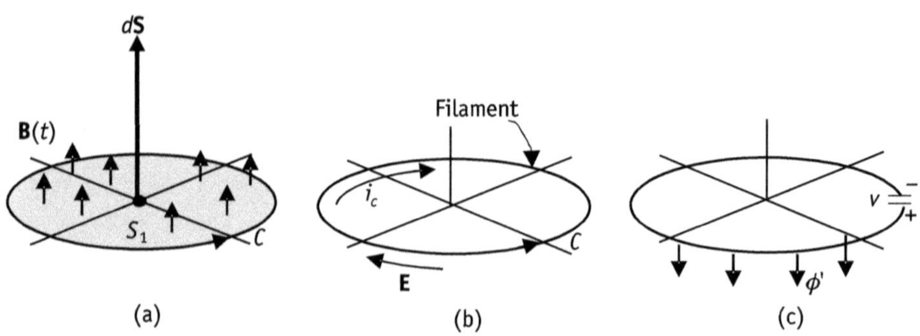

Figure 12.5 (a) Contour C, normal direction $d\mathbf{S}$, (b) direction of $\mathbf{E}$, and (c) current loop generates a flux of ϕ'.

12.4 CONDUCTORS IN MOTION THROUGH TIME-INDEPENDENT FIELDS

The force **F** on a charge Q in a magnetic field **B**, where the charge is moving with velocity **U**, was examined in Chapter 10.

$$\mathbf{F} = Q(\mathbf{U} \times \mathbf{B}) \tag{12.14}$$

A *motional* electric field intensity, $\mathbf{E}_m$, can be defined as the force per unit charge:

$$\mathbf{E}_m = \frac{\mathbf{F}}{Q} = \mathbf{U} \times \mathbf{B} \tag{12.15}$$

When a conductor with a great number of free charges moves through a field, **B**, the impressed $\mathbf{E}_m$ creates a voltage difference between the two ends of the conductor. The magnitude depends on how $\mathbf{E}_m$ is oriented with respect to the conductor. With conductor ends a and b, the voltage of a with respect to b is:

$$v_{ab} = \int_b^a \mathbf{E}_m \cdot d\mathbf{l} = \int_b^a (\mathbf{U} \times \mathbf{B}) \cdot d\mathbf{l} \tag{12.16}$$

If the velocity **U** and the field **B** are at right angles and the conductor is normal to both, then a conductor of length l will have a voltage:

$$v = BlU \tag{12.17}$$

For a closed loop, the line integral must be taken around the entire loop:

$$v = \oint (\mathbf{U} \times \mathbf{B}) \cdot d\mathbf{l} \tag{12.18}$$

Of course, if only part of the complete loop is in motion, it is necessary only that the integral cover this part since $\mathbf{E}_m$ will be zero elsewhere [3].

12.5 CONDUCTORS IN MOTION THROUGH TIME-DEPENDENT FIELDS

When a closed conducting loop is in motion (this includes for loop shape changes) and also the field **B** is a function of time (as well as of position), then the total induced voltage is made up of a contribution from each of the two sources of flux change. Faraday's law becomes:

$$v = -\frac{d}{dt}\int \mathbf{B}\cdot d\mathbf{S} = -\int \frac{\partial \mathbf{B}}{\partial t}\cdot d\mathbf{S} + \oint (\mathbf{U}\times\mathbf{B})\cdot d\mathbf{l} \qquad (12.19)$$

The first term on the right is the voltage due to the change in **B**, with the loop held fixed. The second term is the voltage arising from the motion of the loop with **B** held fixed. The polarity of each term is found from the appropriate form of Lenz's law. The two terms are then added with regard to those polarities [4].

12.6 PROBLEMS AND SOLUTIONS[1]

Problem 12.1

Show that the displacement current, i_D in the dielectric of a parallel plate capacitor, shown in Figure 12.6, is equal to the conduction current, i_c, in the leads.

Problem 12.2

In a material for which $\sigma = 5.0$ S/m and $\varepsilon_r = 1$, the electric field intensity is $E = 250 \sin(10^{10}t)$ V/m.

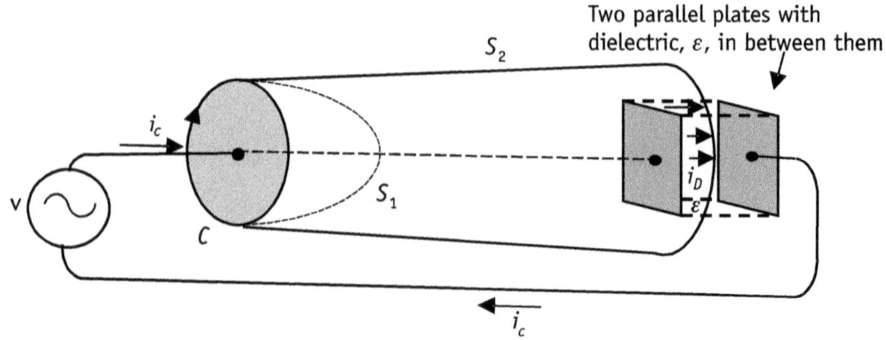

Figure 12.6 Illustration for Problem 12.1.

---

1. See [2].

Find the conduction and displacement current densities and the frequency at which they have equal magnitudes.

Problem 12.3

A coaxial capacitor with an inner radius of 5 mm, outer radius of 6 mm, and a length of 500 mm has a dielectric $\varepsilon_r = 6.7$ and an applied voltage of 250 sin(377t) volts.

Determine the displacement current, i_D. Compare with the conduction current, i_C.

Problem 12.4

In Figure 12.7, a 3m long conductor, parallel to the x-axis, moves with a velocity of $\mathbf{U} = 2.5\mathbf{y}$ m/s while in a uniform field of $\mathbf{B} = 0.5\mathbf{z}$ tesla.

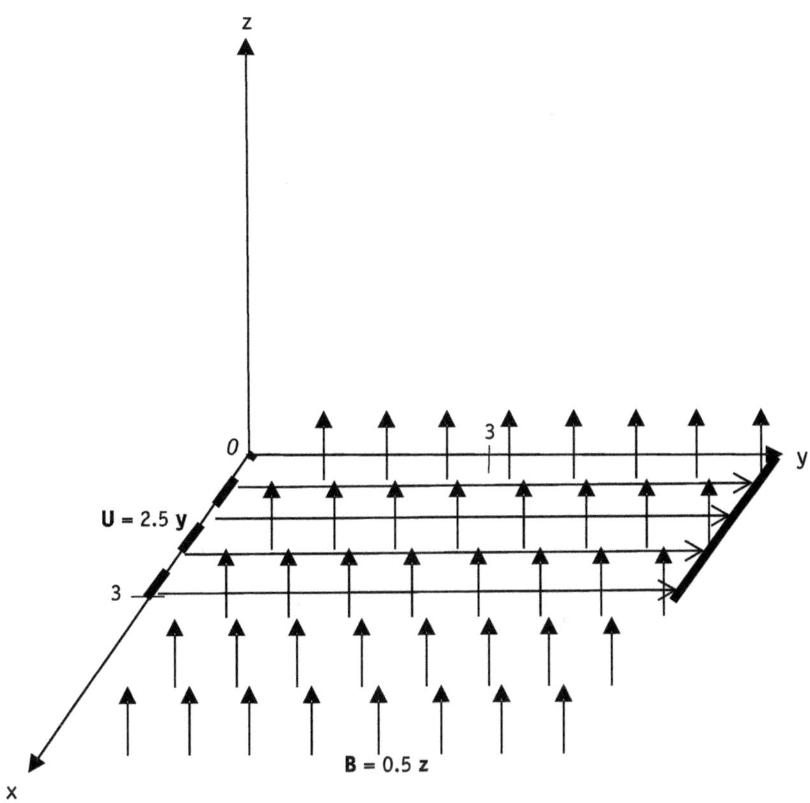

Figure 12.7 Illustration for Problem 12.4.

Find the induced voltage.

Problem 12.5

With the same field and velocity as given in the previous problem, find the voltage induced in the same 3m conductor lying along the *y*-axis. See Figure 12.8.

Problem 12.6

Find the induced voltage in the conductor in Figure 12.9 where $\mathbf{B} = 0.04\mathbf{y}$ tesla and $\mathbf{U} = 2.5 \sin(10^3\, t)\mathbf{z}$ (m/s).

Problem 12.7

An area of 0.65m^2 in the $z = 0$ plane is enclosed by a filamentary conductor.

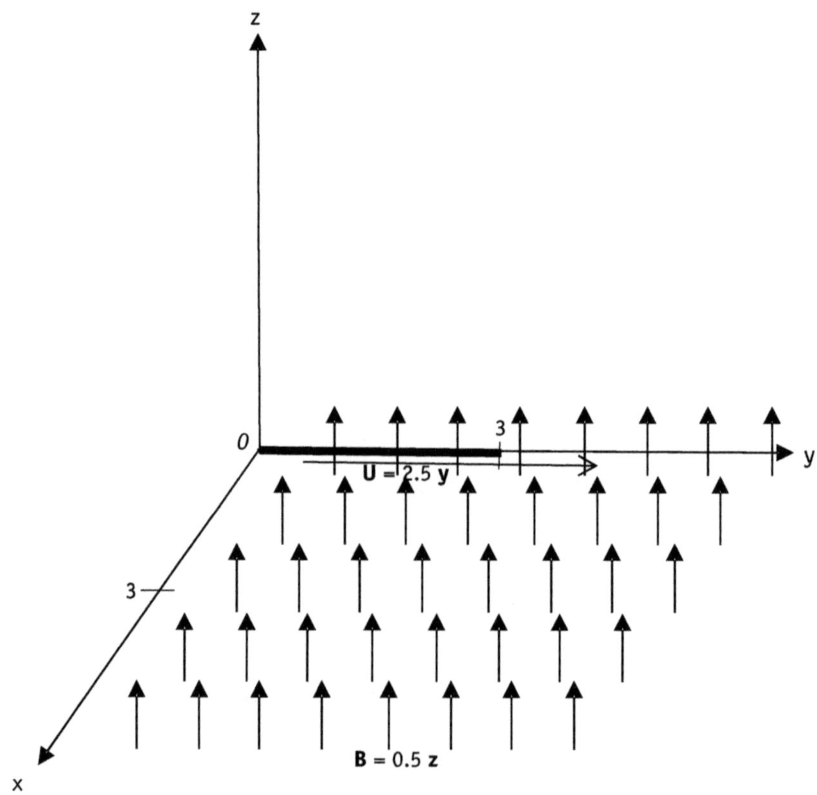

Figure 12.8 Illustration for Problem 12.5.

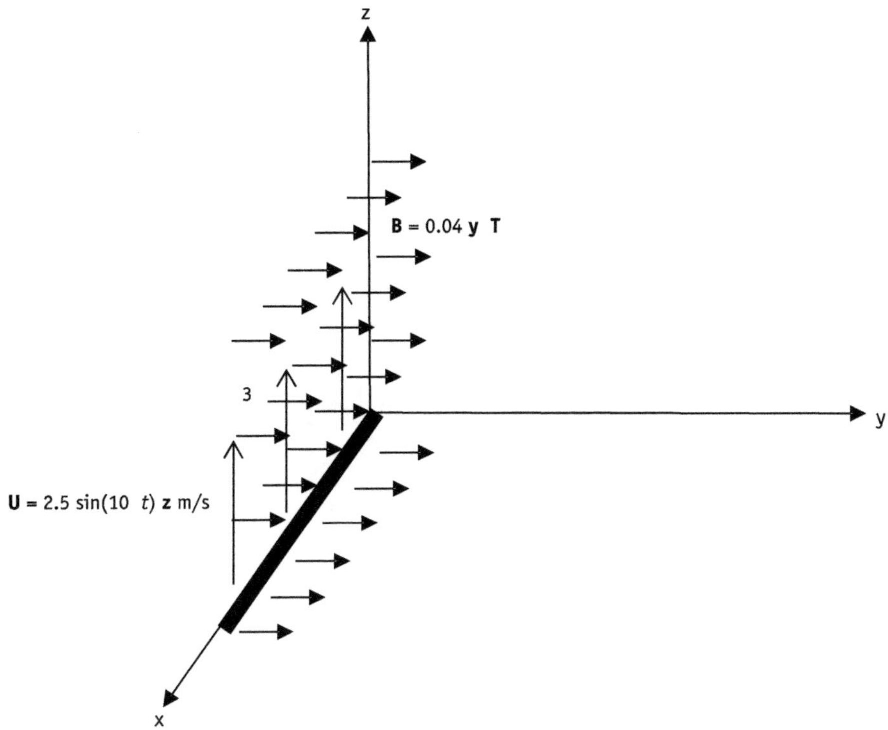

Figure 12.9 Illustration for Problem 12.6.

Find the induced voltage, given the field of:

$$\mathbf{B} = [(0.05/2^{1/2})\cos(10^3 t)]\ (\mathbf{y} + \mathbf{z})\ \text{tesla}$$

See Figure 12.10.

Problem 12.8

The circular loop conductor shown in Figure 12.11 lies in the $z = 0$ plane, has a 0.1m radius, and has a resistance of 5.0Ω.

Given $\mathbf{B} = (0.20)\sin(10^3 t)\mathbf{z}$ tesla, determine the current in the loop.

Problem 12.9

The rectangular loop shown in Figure 12.12 moves toward the origin at a velocity $\mathbf{U} = -250\mathbf{y}$ m/s in a field given by

$$\mathbf{B} = (0.80)\ e^{-0.5y}\mathbf{z}\ \text{tesla}$$

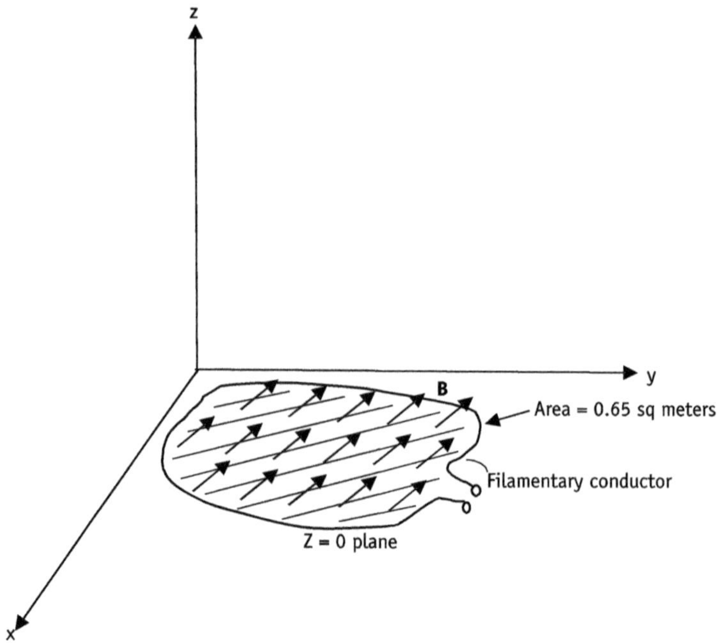

Figure 12.10 Illustration for Problem 12.7.

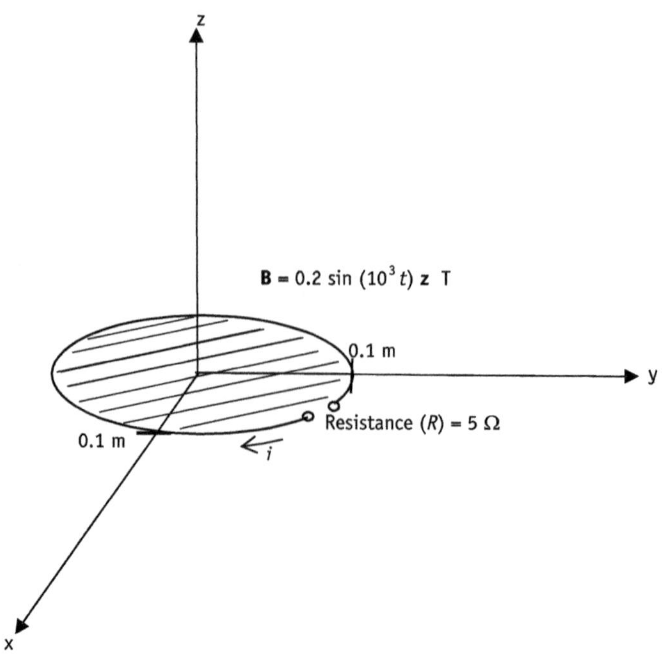

Figure 12.11 Illustration for Problem 12.8.

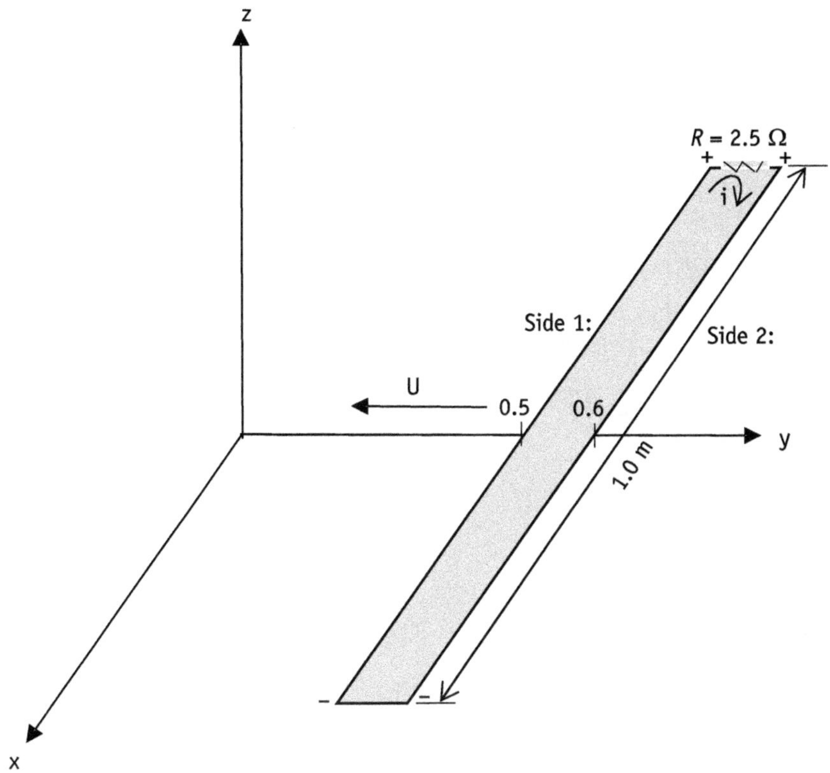

Figure 12.12 Illustration for Problem 12.9.

Find the current at the instant the coil sides are at $y = 0.5$m and 0.6m if $R = 2.5\Omega$.

Problem 12.10

A conducting cylinder of radius 7 cm and height of 15 cm rotates at 600 rpm in a radial field given by $\mathbf{B} = (0.2)\mathbf{r}$ tesla. Sliding contacts at the top and bottom connect to a voltmeter as shown in Figure 12.13.

Find the induced voltage.

Problem 12.11

In Figure 12.14, a rectangular conducting loop with resistance of $R = 0.2\Omega$ turns at a rate of 500 rpm. The vertical conductor a $r_1 = 0.03$m is in a field described as

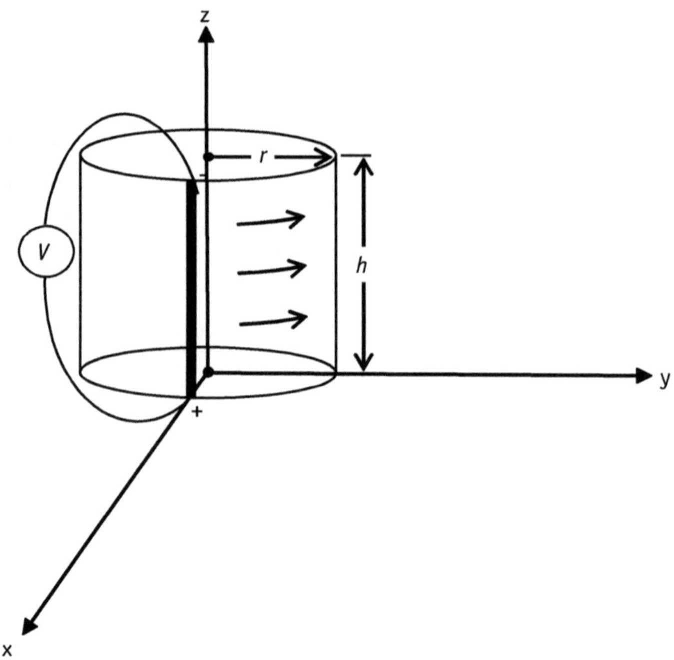

Figure 12.13 Illustration for Problem 12.10.

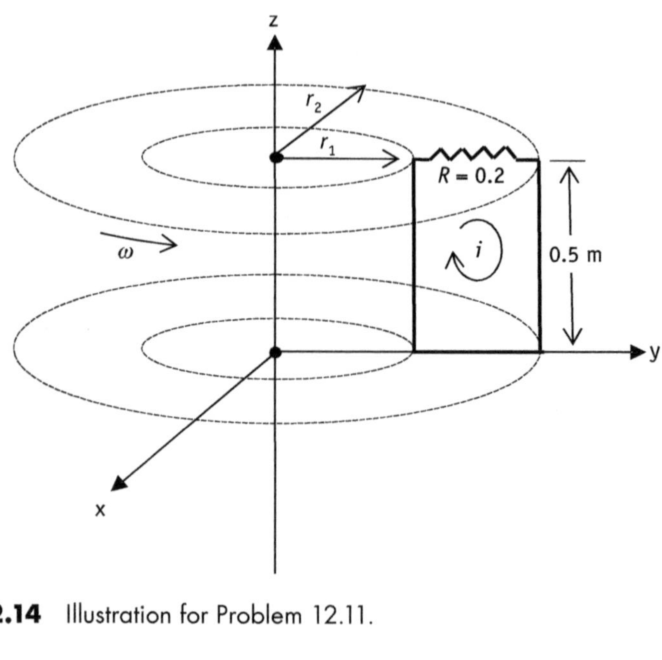

Figure 12.14 Illustration for Problem 12.11.

$$\mathbf{B} = (0.25)\mathbf{r} \text{ tesla}$$

The conductor at $r_2 = 0.05$m is in a field described as

$$\mathbf{B} = (0.8)\mathbf{r} \text{ tesla}$$

Find the current in the loop.

References

[1] Russer, P., *Electromagnetics, Microwave Circuit and Antenna Design for Communications Engineering*, Norwood, MA: Artech House, 2006.

[2] Shadowitz, A., *The Electromagnetic Field*, New York: Dover Publications, 1975.

[3] Warnick, K. F., and P. Russer, *Problem Solving in Electromagnetics, Microwave Circuit and Antenna Design for Communications Engineering*, Norwood, MA: Artech House, 2006.

[4] Cheng, D. K., *Fundamentals of Engineering Electromagnetics*, Reading, MA: Addison-Wesley, 1993.

PART 3

PRINCIPLES OF FREE-SPACE ELECTROMAGNETICS

13

ELECTROMAGNETIC WAVES

This chapter contains an elementary discussion of EM wave theory.

13.1 CHARACTERISTICS OF ELECTROMAGNETIC WAVES

The focus of this section is to provide an elementary discussion of electromagnetic waves, and the aim has been to make it essentially free of an onslaught of equations. A first conceptual understanding of RF/MW/mm frequency waves is attempted as follows.

Wave Velocity

EM waves oscillate transversely, unlike sound waves, which oscillate longitudinally. In empty free space EM waves travel at a speed of approximately 300×10^6 m/s (i.e., more precisely, 299,792,458 m/s). However, in air, other gases, or other dielectric media, EM waves travel more slowly, the resulting speed depending entirely upon the permittivity, ε, of the dielectric media through which the wave is passing relative to free space [8].

Frequency and Wavelength

EM wave oscillations are also periodic. Furthermore, single frequency waves are sinusoidal as opposed to more complex periodic waves that are composed of more than one sinusoidal frequency component. The frequency of the oscillating wave is measured in hertz (Hz or cycles per second) with the length of a purely sinusoidal wave being measured in meters as determined by the following formula:

$$\lambda = \left(\frac{c}{f}\right) \tag{13.1}$$

where c = speed of light (m/s) and f = frequency (Hz) in free, empty space.

Space-Time Relationships

An EM wave consists of two components (i.e., an electric (or E-) field and a magnetic (or H-) field). The fields are orthogonal (i.e., perpendicular) to one another, and travel together in a direction that is also orthogonal to each field (see Figure 13.1.) [1].

While traveling in lossless free space, the E- and H-fields of the EM wave are in phase to each other; that is, phase angle between them being 0 radians (or 0 degrees). The fields vary both sinusoidally and simultaneously in time at a fixed point in free space. They also have a periodicity, T, as can be determined by the formula:

$$T = \frac{1}{f} \tag{13.2}$$

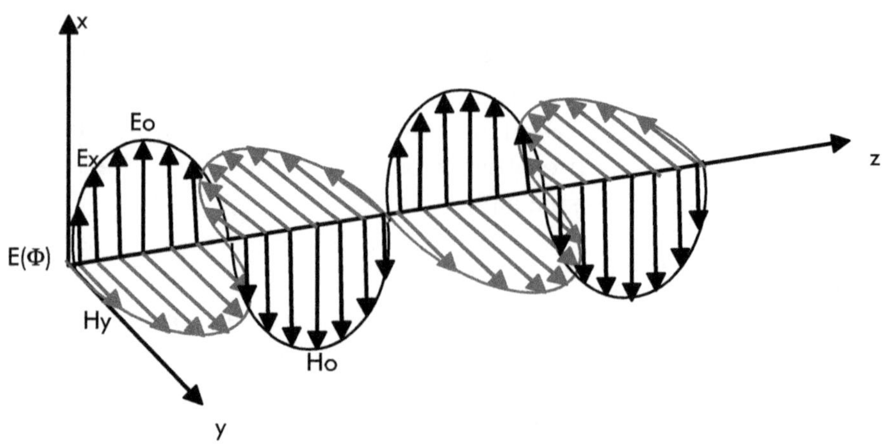

Figure 13.1 Far-field plane wave as represented in the Cartesian (x, y, z) coordinate system [1].

where f = frequency (Hz) and T = period (in seconds).

When described in terms of the Cartesian (x,y,z) coordinate system, the E-field is represented as parallel to the x-axis and the H-field is represented as parallel to the y-axis. The wave's travel direction will be with respect to the z-axis as shown in Figure 13.1. Time-domain equations for the E-field and H-field [2], respectively, are

$$E_x(z,t) = E_o \sin\left[2\pi ft + \varphi - 2\pi z/\lambda\right] \quad (\text{V/m}) \tag{13.3}$$

E_o = the waveform maximum value (V/m)

$\pi = 3.1415927$

f = operating frequency of waveform (Hz)

t = time (s)

φ = waveform phase shift (radians)

z = physical displacement along z-axis (m)

λ = wavelength (m)

$$H_y(z,t) = H_o \sin\left[2\pi ft + \varphi - 2\pi z/\lambda\right] \quad (\text{A/m}) \tag{13.4}$$

H_o = the waveform maximum value (A/m)

$\pi = 3.1415927. \ldots$

f = operating frequency of waveform (Hz)

t = time (s)

φ = waveform phase shift (radians)

z = physical displacement along z-axis (m)

λ = wavelength (m)

with the following relation to help find the peak values of one quantity knowing the peak value of the other (note that this is in free space wave propagation conditions only):

$$H_o = \frac{E_o}{Z_o} = \frac{E_o}{120\pi} \tag{13.5}$$

$Z_o = 120\pi\ \Omega$ = characteristic impedance of free space

Thus, in free space the E-field and H-field are always in phase and at right angles to each other as each travels in the x-y plane normal to and along the z-axis [8].

Far-Field Versus Near-Field

While we speak ideally of plane waves propagating, the truth is that the wave fronts are never perfectly planar. One should note that the farther away from the source that an EM wave has traveled, the more increasingly planar (or flat) the wave becomes (see Figure 13.2) [2].

Thus, it follows logically to ask the question: "At what distance is a propagating EM wave, for all intents and purposes, considered to be planar?" This boundary is not sharply defined.

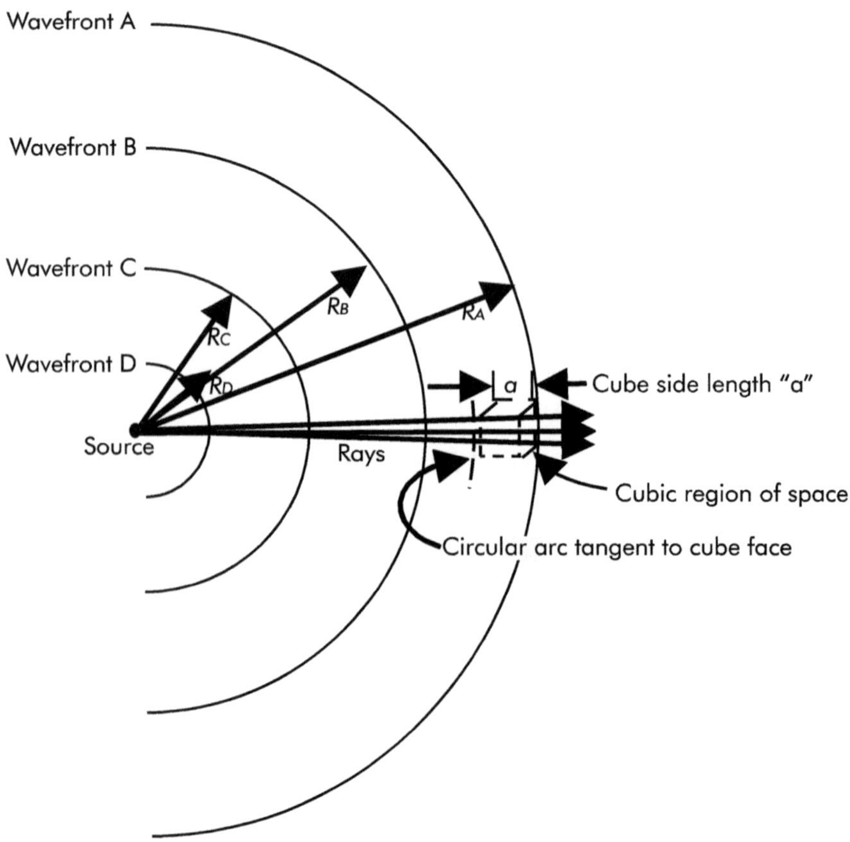

Figure 13.2 Point-source wavefronts and rays in free space.

13.1.1 Radiating Far-Field Region Versus Radiating Near-Field Region

Starting from a very long distance away from the antenna and moving physically closer to it, you eventually begin to move out of the antenna's radiating far-field (Fraunhofer) region and into its radiating near-field (Fresnel) region. Decades ago, this boundary was debated with technical paper discussions advocating for both the D^2/λ and the $4D^2/\lambda$ equations. Strong arguments were made for both positions. In the end, antenna experts settled on $2D^2/\lambda$. Suffice it to say that $2D^2/\lambda$ has worked well ever since. The author has used this equation successfully for 30 years of his career.

In addition, in 1989 the author got to work an entire year with the late Dr. James McDougal at the Boeing Company in Wichita, KS (Dr. McDougal retired at the end of the year). Dr. McDougal taught him the derivation of the near-field/far-field distance formula. (Note: The distance is measured away from the aperture antenna face.)

Referencing Figure 13.3, the boundary formula may also be expressed as:

$$R_{FF} > \frac{2D^2}{\lambda} \tag{13.6a}$$

$$R_{FF} = \frac{2D^2}{\lambda} \tag{13.6b}$$

and

$$R_{NF} < \frac{2D^2}{\lambda} \tag{13.7}$$

D = longest antenna dimension (m)

λ = wavelength (m).

A complete step-by-step derivation for expression (13.6b) may be found in [7].

By settled antenna engineering convention, the formula in (13.6b) is the minimum far-field distance between a radiating point source and a receive antenna aperture of dimension D. This convention has

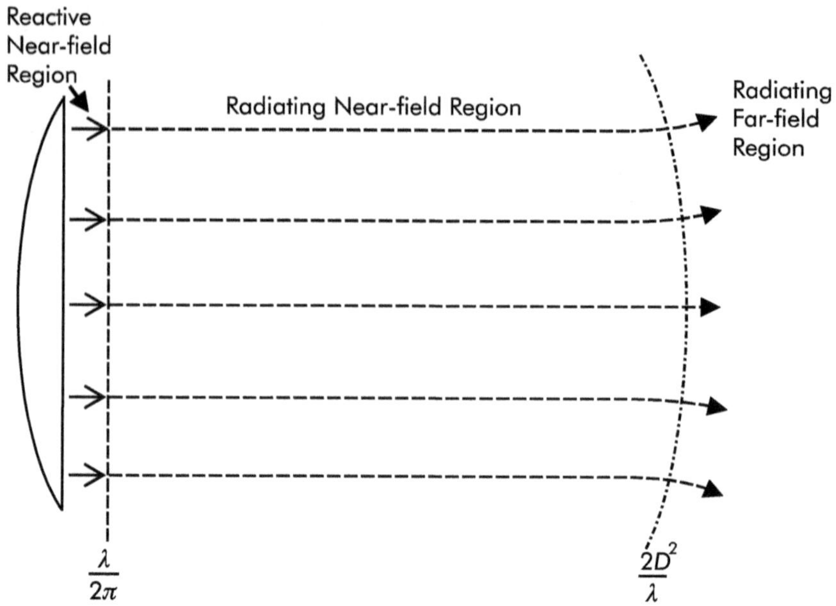

Figure 13.3 Definition for the reactive near-field equation for aperture antennas.

remained stable for more than 50 years as of the date of this book release. As for the near-field region, there is more to know.

13.1.2 Reactive Near-Field Region Versus Radiating Near-Field Region

Within the near-field of an antenna, the accepted boundary for the reactive and radiating near-field regions is conventionally defined as

$$R = \frac{\lambda}{2\pi} \tag{13.8}$$

λ = wavelength (m).

Also, see Figure 13.3.

Simply put, the *reactive near-field region* is the region immediately surrounding the antenna where the reactive field predominates. The E-field and H-field are 90 degrees out of phase with each other, and this is the region where the radiating fields begin to emerge.

Polarization

By accepted convention, the polarization of an EM wave is always determined with respect to the direction of the plane wave's E-field. The far-field EM plane wave illustrated in Figure 13.1 is *linearly* polarized in the vertical (or *x*-axis) direction, that is, with respect to the Earth's "up/down" reference. However, by rotating the launching antenna 90 degrees from the vertical reference position, either clockwise or counterclockwise in the *x-y* plane, the EM wave can be polarized horizontally [8].

It is also possible to polarize an EM wave circularly, that is, cause the wave electromagnetically to travel forward through free space with the orthogonal E/H-fields rotating in either a clockwise (right-hand circular or RHC) or counterclockwise (left-hand circular or LHC) direction. By accepted convention, the direction of E-/H-field rotation is determined from the perspective of the fixed position observer watching the circularly polarized plane wave advance toward him/her [2].

Rays and Wavefronts

A *wavefront* is defined as a wave surface having a constant phase difference between the E-field and H-field. A *ray*, on the other hand, is a line that is drawn to show the direction of EM wave propagation in space (i.e., the *z*-axis as shown in Figure 13.1) [2].

Having now briefly discussed the basic characteristics of free-space EM wave propagation, the author recommends the reader to procure the following highly-rated (5-star out of 5, Amazon) Artech House technical book [7]. Reference [7] was made available to the general public in 2022. It was written to show a new generation of RF/MW engineers how to analyze and numerically evaluate the value(s) for the expected received RF power as referenced at the front-end of an RF receiver system, that is, when a RF transmit system propagates an EM wave a long distance (i.e., in the far-field) away from the EM wave receiving system. The information contained in [1] is relevant to EM wave propagation between low frequencies (150 MHz) and millimeter-wave frequencies (94 GHz) [8]. Optical (or above, at, or below visible light) wavelengths are not addressed in the cited reference or in this edition.

In the next section, the subject of electromagnetic wave principles continues.

13.2 PRINCIPLES OF ELECTROMAGNETIC WAVES

This section gives a fuller, more complete treatment of the subject of electromagnetic waves. Since most regions of interest are free of charge, it will be assumed that $\rho = 0$, and, moreover, linear isotropic materials will be assumed such that [6]

$$\mathbf{D} = \varepsilon \mathbf{E} \tag{13.9}$$

$$\mathbf{B} = \mu \mathbf{H} \tag{13.10}$$

$$\mathbf{J}_c = \sigma \mathbf{E} \tag{13.11}$$

Wave Equations

Under the above assumptions and supposing that both **E** and **H** have the time-dependence $e^{j\omega t}$, Maxwell's equations give [6]:

$$\nabla \times \mathbf{H} = (\sigma + j\omega\varepsilon)\mathbf{E} \tag{13.12}$$

$$\nabla \times \mathbf{E} = -j\omega\mu \mathbf{H} \tag{13.13}$$

$$\nabla \cdot \mathbf{E} = 0 \tag{13.14}$$

$$\nabla \cdot \mathbf{H} = 0 \tag{13.15}$$

At this point, apply the vector identity

$$\nabla \times (\nabla \times \mathbf{A}) = \nabla(\nabla \cdot \mathbf{A}) - \nabla^2 \mathbf{A} \tag{13.16}$$

where, in Cartesian coordinates only,

$$\nabla^2 \mathbf{A} = (\nabla^2 A_x)\mathbf{a}_x + (\nabla^2 A_y)\mathbf{a}_y + (\nabla^2 A_z)\mathbf{a}_z \tag{13.17}$$

Taking the curl of (13.12) and (13.13) and using (1.14) and (1.15) yields the following,

$$-\nabla^2\mathbf{H} = (\sigma + j\omega\varepsilon)(\nabla \times \mathbf{E}) \tag{13.18}$$

$$-\nabla^2\mathbf{E} = -j\omega\mu(\nabla \times \mathbf{H}) \tag{13.19}$$

Substituting $\nabla \times \mathbf{E}$ and $\nabla \times \mathbf{H}$ from (13.13) and (13.12), one obtains the vector *wave equations*:

$$\nabla^2\mathbf{H} = \gamma^2\mathbf{H} \tag{13.20}$$

$$\nabla^2\mathbf{E} = \gamma^2\mathbf{E} \tag{13.21}$$

where $\gamma^2 = j\omega\mu\ (\sigma + j\omega\varepsilon)$.

The *propagation constant*, γ, is that square root of γ^2 whose real and imaginary parts are positive:

$$\gamma = \alpha + j\beta \tag{13.22}$$

$$\alpha = \omega\{½\mu\varepsilon(1 + (\sigma/\omega\varepsilon)^{½} - 1\}^{½} \quad \textit{(attenuation factor)} \tag{13.23}$$

$$\beta = \omega\{½\mu\varepsilon(1 + (\sigma/\omega\varepsilon)^{½} + 1\}^{½} \quad \textit{(phase shift constant)} \tag{13.24}$$

The constant α is called the *attenuation factor*; β is called the *phase shift constant*. The reason for these names will become apparent in Section 14.4. From the wave equations, it is seen that γ carries the units m^{-1}. However, it is customary to give α in Np/m and β in rad/m, respectively, where the *neper* (Np) is a dimensionless unit like the radian [5].

Solutions in Cartesian Coordinates

The familiar scalar wave equation in one dimension,

$$\frac{\partial^2 F}{\partial z^2} = \left(\frac{1}{U^2}\right)\frac{\partial^2 F}{\partial t^2} \tag{13.25}$$

has solutions of the form

$$F = f(z - Ut) \tag{13.26}$$

and

$$F = g(z + Ut) \tag{13.27}$$

where f and g are arbitrary functions. These represent waves traveling with speed U and in the $+z$ and $-z$ directions, respectively. See Figure 13.4.

In particular, if harmonic time variation $e^{j\omega t}$ is assumed, the wave equations become

$$\frac{d^2F}{dz^2} = -\beta^2 F \tag{13.28a}$$

$$\beta = \frac{\omega}{U} \tag{13.28b}$$

with solutions (including the time factor) shown below being of the form

$$F = Ce^{j(\omega t - \beta z)} \tag{13.29}$$

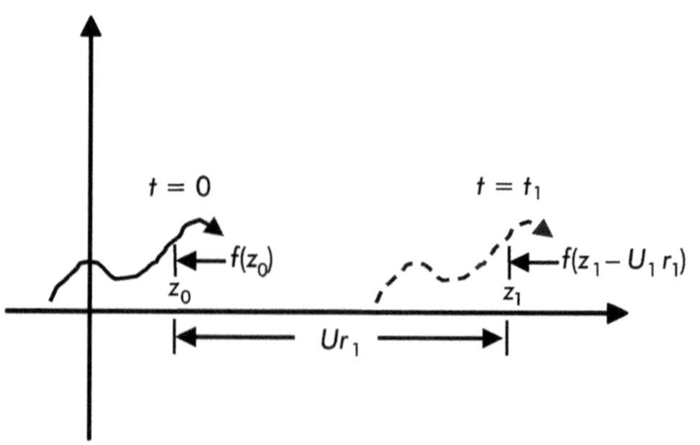

Figure 13.4 Waves traveling with the speed of U in the $+z$ and $-z$ directions.

and

$$F = De^{j(\omega t+\beta z)} \tag{13.30}$$

or the real or imaginary parts of these.

During this time interval the wave has moved to the right a distance of

$$d = U\left(\frac{\pi}{2\omega}\right) = \frac{\pi}{2\beta} \tag{13.31}$$

At any fixed t, the waveform repeats itself when z changes by $2\pi/\beta$. The distance:

$$\lambda = \frac{2\pi}{\beta} \tag{13.32}$$

is called the *wavelength*. The wavelength and the *frequency*,

$$f = \frac{\omega}{2\pi} \tag{13.33}$$

have the well-known relation

$$U = \lambda f \tag{13.34}$$

Alternatively,

$$\lambda = TU \tag{13.35a}$$

$$T = \frac{2\pi}{\omega} \tag{13.35b}$$

where T is called the *period*.

Because the unit vectors $\mathbf{a}_x$, $\mathbf{a}_y$, and $\mathbf{a}_z$ in Cartesian coordinates have fixed directions, the wave equation for $\mathbf{H}$ can be rewritten in the form:

$$\frac{\partial^2 \mathbf{H}}{\partial x^2} + \frac{\partial^2 \mathbf{H}}{\partial y^2} + \frac{\partial^2 \mathbf{H}}{\partial z^2} = \gamma^2 \mathbf{H} \tag{13.36}$$

Of particular interest are solutions (plane waves) that depend on only one spatial coordinate, say, z. Then, the equation becomes:

$$\frac{d^2 \mathbf{H}}{dz^2} = \gamma^2 \mathbf{H} \tag{13.37}$$

yielding

$$\mathbf{H} = H_0 e^{\pm \gamma z} \mathbf{a}_H \tag{13.38}$$

or

$$\mathbf{H}(z,t) = H_0 e^{\pm \gamma z} e^{j\omega t} \mathbf{a}_H \tag{13.39}$$

The corresponding electric field solutions are:

$$\mathbf{E} = E_0 e^{\pm \gamma z} \mathbf{a}_E \tag{13.40}$$

or

$$\mathbf{E}(z,t) = E_0 e^{\pm \gamma z} e^{j\omega t} \mathbf{a}_E \tag{13.41}$$

Here, $\mathbf{a}_H$ and $\mathbf{a}_E$ are fixed unit vectors [1]. The complex quantity γ is as given in Section 13.2. It should be noted that it may also be shown that

$$\mathbf{a}_H \cdot \mathbf{a}_z = \mathbf{a}_E \cdot \mathbf{a}_z = 0 \tag{13.42}$$

That is, neither field has a component in the direction of propagation. This being the case, one can always rotate the axes to put one of the fields, say **E**, along the x-axis. **H** lies along the y-axis. The plane wave solutions obtained depend, via γ, on the properties of the medium (i.e., μ, ε, σ) [5]. This dependence will be examined next. Solutions for partially conducting media (lossy dielectric materials), perfect (ideal) dielectric materials, and good electrically conductive materials are presented as follows.

Partially (Electrically) Conducting Media

For a region in which there is some conductivity but not much (e.g., moist earth, seawater), the solution to the wave equation in **E** is taken to be:

$$\mathbf{E} = E_0 e^{-\gamma z} \mathbf{a}_x \tag{13.43}$$

Then, from (13.13) of Section 13.2:

$$\mathbf{H} = \left[\frac{(\sigma + j\omega\varepsilon)}{j\omega\mu} \right]^{1/2} E_0 e^{-\gamma z} \mathbf{a}_y \tag{13.44}$$

The ratio E/H is characteristic of the medium (it is also frequency-dependent). More specifically, for waves

$$\mathbf{E} = E_x \mathbf{a}_x \tag{13.45}$$

and

$$\mathbf{H} = H_y \mathbf{a}_y \tag{13.46}$$

Which propagate in the $+z$ direction, the *intrinsic impedance*, η, of the medium is defined by:

$$\eta = \frac{E_x}{H_y} \tag{13.47}$$

Thus,

$$\eta = \left[\frac{j\omega\mu}{(\sigma + j\omega\varepsilon)}\right]^{1/2} \tag{13.48}$$

where the correct square root may be written in the polar form, $|\eta| < \theta$, with

$$|\eta| = \frac{\left(\frac{\mu}{\varepsilon}\right)^{1/2}}{\left[1 + \left(\frac{\sigma}{\omega\varepsilon}\right)^2\right]^{1/4}} \tag{13.49}$$

$$\tan 2\theta = \frac{\sigma}{\omega\varepsilon} \quad \text{where: } 0° < \theta < 45° \tag{13.50}$$

(If the wave propagates in the $-z$ direction, then

$$\frac{E_x}{H_y} = -\eta \tag{13.51}$$

in effect, γ is replaced by $-\gamma$ and the other square root is used.)

Inserting the time factor, $e^{j\omega t}$, and writing

$$\gamma = \alpha + j\beta \tag{13.52}$$

results in the following equations for fields in a partially conducting region:

$$\mathbf{E}(z,t) = E_0 e^{-\alpha z} e^{j(\omega t - \beta z)} \mathbf{a}_x \tag{13.53}$$

$$\mathbf{H}(z, t) = \left[\frac{E_0}{|\eta|}\right] e^{-\alpha z} e^{j(\omega t - \beta z - \theta)} \mathbf{a}_y \tag{13.54}$$

The factor $e^{-\alpha z}$ attenuates the magnitudes of both **E** and **H** as they propagate in the +z direction. The expression for α, (13.23), shows that there will be some attenuation unless the conductivity σ is zero, which would be the case only for perfect dielectrics or free space. Likewise, the time phase difference θ between $\mathbf{E}(z,t)$ and $\mathbf{H}(z,t)$ vanishes only when σ is zero [5].

The velocity of propagation and the wavelength are given by:

$$U = \omega/\beta = 1/\{½\mu\varepsilon[(1 + (\sigma/\omega\varepsilon)^2)^{½} + 1\}]^{½} \quad \text{(m/s)} \tag{13.55}$$

$$\lambda = 2\pi/\beta = 2\pi/[\omega\{½\mu\varepsilon[(1 + (\sigma/\omega\varepsilon)^2)^{½} + 1]\}^{½}] \quad \text{(m)} \tag{13.56}$$

If the propagation velocity is known, $\lambda f = U$ may be used to determine the wavelength λ. The term $(\sigma/\omega\varepsilon)^2$ has the effect of reducing both the velocity and the wavelength from what they would be either free space or perfect dielectric materials where $\sigma = 0$. Observe that the medium is dispersive; that is, waves with different frequencies ω have different velocities, U [1].

Perfect Dielectrics. For a perfect dielectric [3],

$$\sigma = 0 \tag{13.57}$$

and so:

$$\alpha = 0 \tag{13.58}$$

$$\beta = \omega(\mu\varepsilon)^{1/2} \quad \text{and} \tag{13.59}$$

$$\eta = (\mu/\varepsilon)^{½} < 0 \text{ deg} \tag{13.60}$$

Since $\alpha = 0$, there is no attenuation of the **E** and **H** waves. The zero angle on η results in **H** being in time phase with **E** at each fixed location. Assuming **E** in $\mathbf{a}_x$ and propagating in $\mathbf{a}_z$, the field equations may be obtained as limits of those in the previous section shown as follows:

$$\mathbf{E}(z,t) = E_0 e^{j(\omega t - \beta z)} \mathbf{a}_x \tag{13.61}$$

$$\mathbf{H}(z,t) = \frac{E_0}{\eta} e^{j(\omega t - \beta z)} \mathbf{a}_y \tag{13.62}$$

The velocity and the wavelength are:

$$U = \frac{\omega}{\beta} = \frac{1}{(\mu\varepsilon)^{1/2}} \tag{13.63}$$

$$\lambda = \frac{2\pi}{\beta} = \frac{2\pi}{\left[\omega(\mu\varepsilon)^{1/2}\right]} \tag{13.64}$$

Free Space

Free space is nothing more than the perfect dielectric for which:

$$\mu = \mu_0 = 4\pi \times 10^{-7} \ \text{H/m} \tag{13.65}$$

$$\varepsilon = \varepsilon_0 = 8.854 \times 10^{-12} \ \text{F/m} \tag{13.66}$$

For free space,

$$\eta = \eta_0 \sim 120\pi \ \Omega \tag{13.67}$$

and

$$U = c \sim 3 \times 10^8 \ \text{m/s} \tag{13.68}$$

Good Conductors

In the ordinary materials classified as conductors,

$$\sigma >> \omega\varepsilon \tag{13.69}$$

in the range of practical frequencies. For example, in copper, where $\sigma = 5.80 \times 10^7$ S/m and $\varepsilon \simeq \varepsilon_0 = 8.854 \times 10^{-12}$ F/m, it would be necessary for frequencies to be of the order of 10^{16} Hz before $\omega\varepsilon/\sigma$ would have to be retained in the expressions for the propaganda constant and the intrinsic impedance. Therefore, letting $\omega\varepsilon/\sigma \to 0$.

$$\alpha = \beta = \left(\frac{1}{2}\omega\mu\sigma\right)^{1/2} \tag{13.70}$$

$$\eta = \left(\frac{\omega\mu}{\sigma}\right)^{1/2} < 45^\circ \tag{13.71}$$

It is seen that for all conductors the **E** and **H** waves are attenuated. Numerical examples will show that this is a very rapid attenuation. α will always be equal to β. At each fixed location, **H** is out of time phase with **E** by 45° or $\pi/4$ rad. Once again assuming **E** in $\mathbf{a}_x$ and propagation in $\mathbf{a}_z$, the field equations are,

$$\mathbf{E}(z,t) = E_0 e^{-\alpha z} e^{j(\omega t - \beta z)} \mathbf{a}_x \tag{13.72}$$

$$\mathbf{H}(z,t) = \frac{E_0}{|\eta|} e^{-\alpha z} e^{j(\omega t - \beta z - \pi/4)} \mathbf{a}_y \tag{13.73}$$

Moreover,

$$U = \frac{\omega}{\beta} = \left(\frac{2\omega}{\mu\sigma}\right)^{1/2} = \omega\delta \tag{13.74}$$

$$\lambda = \frac{2\pi}{\beta} = \frac{2\pi}{(\pi f \mu\sigma)^{1/2}} = 2\pi\delta \tag{13.75}$$

The velocity and wavelength in a conducting medium are written here in terms of the *skin depth* or *depth of penetration*,

$$\delta \simeq \frac{1}{(\pi f \mu\sigma)^{1/2}} \tag{13.76}$$

Skin Depth

The **E** and **H** traveling waves in a conducting medium are attenuated by the factor $e^{-\alpha z}$ as they advance along z [3]. This attenuation is so rapid that often the waves may be considered zero after only a few millimeters of travel in that medium.

Consider that the region $z \geq 0$ is a conductor and just inside the conductor at $z = +0$, **E** has magnitude of 1.0 V/m. The *skin depth*, δ, is defined to be the distance after which $|\mathbf{E}|$ has decreased to:

$$e^{-1} = 0.368 \ \text{V/m} \quad [3] \tag{13.77}$$

See Figure 13.5.

Thus,

$$\delta = \frac{1}{\alpha} = \frac{1}{(\pi f \mu \sigma)^{1/2}} \ (\text{m}) \tag{13.78}$$

For convenience, $z = 5\delta$ is often taken as the point where the function is zero since the value there is 0.0067 or 0.67% of the initial value. For example, at a frequency of 100 MHz in copper, the skin depth is 6.61 μm. The waves are attenuated to 0.67% in 5δ or 33.0 μm. Consequently, the term propagation when used in conjunction with wave behavior within a conductor is somewhat misleading. The **E** and **H** waves hardly propagate at all. As will be determined shortly, the major part of an incident wave at a conductor surface is reflected. However, the portion which continues into the conductor and is rapidly attenuated cannot be completely ignored since it gives rise to a conduction current density, J_c, and attendant ohmic power loss.

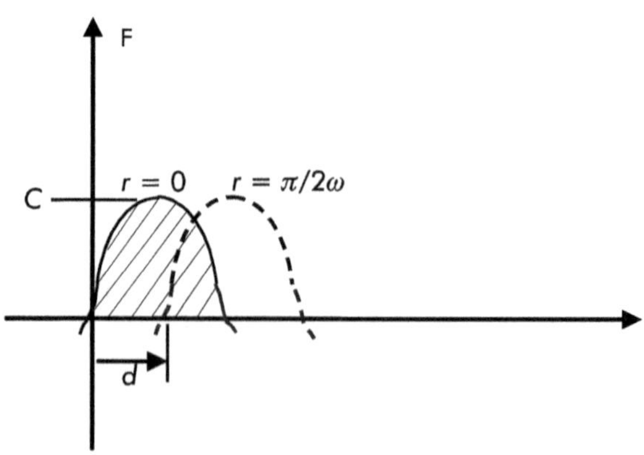

Figure 13.5 Graph of equation 13.77.

Reflected Waves

When a traveling wave reaches an interface between two different regions, it is partly reflected and partly transmitted with the magnitudes of the two parts determined by the constants of the two regions [1]. In Figure 13.6, a traveling **E** wave approaches the interface $z = 0$ from region 1, $z < 0$. $\mathbf{E}^i$ and $\mathbf{E}^r$ are at $z = -0$, while $\mathbf{E}^t$ is at $z = +0$ (in region 2). In this notation, "i" signifies incident, "r" for reflected, and "t" for transmitted. Normal incidence is assumed. The equations for **E** and **H** can be written:

$$\mathbf{E}^i(z,t) = E_0^i e^{-\gamma 1z} e^{j\omega t} \mathbf{a}_x \tag{13.79}$$

$$\mathbf{E}^r(z,t) = E_0^r e^{\gamma 1z} e^{j\omega t} \mathbf{a}_x \tag{13.80}$$

$$\mathbf{E}^t(z,t) = E_0^t e^{-\gamma 2z} e^{j\omega t} \mathbf{a}_x \tag{13.81}$$

$$\mathbf{H}^i(z,t) = H_0^i e^{-\gamma 1z} e^{j\omega t} \mathbf{a}_y \tag{13.82}$$

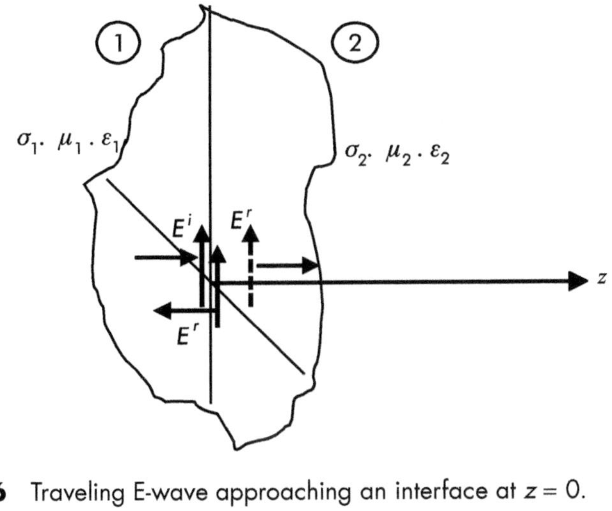

Figure 13.6 Traveling E-wave approaching an interface at $z = 0$.

$$\mathbf{H}^r(z,t) = H_0^r e^{\gamma 1 z} e^{j\omega t}\mathbf{a}_y \tag{13.83}$$

$$\mathbf{H}^t(z,t) = H_0^t e^{-\gamma 2 z} e^{j\omega t}\mathbf{a}_y \tag{13.84}$$

With normal incidence, **E** and **H** are entirely tangential to the interface and, thus, are continuous across it. At $z = 0$ this implies

$$E_0^i + E_0^r = E_0^t \tag{13.85}$$

$$H_0^i + H_0^r = H_0^t \tag{13.86}$$

Furthermore, the intrinsic impedance in either region is equal to $\pm E_x H_y$.

$$\frac{E_0^i}{H_0^i} = \eta_1 \tag{13.87}$$

$$E_0^r / H_0^r = -\eta_1 \tag{13.88}$$

$$\frac{E_0^t}{H_0^t} = \eta_2 \tag{13.89}$$

The five above equations shown can be combined to produce the following ratios in terms of the intrinsic impedances:

$$\frac{E_0^r}{E_0^i} = \frac{(\eta_2 - \eta_1)}{(\eta_1 + \eta_2)} \tag{13.90}$$

$$\frac{H_0^r}{H_0^i} = \frac{(\eta_1 - \eta_2)}{(\eta_1 + \eta_2)} \tag{13.91}$$

$$\frac{E_0^t}{E_0^i} = \frac{2\eta_2}{(\eta_1 + \eta_2)} \tag{13.92}$$

$$\frac{H_0^t}{H_0^i} = \frac{2\eta_2}{(\eta_1 + \eta_2)} \tag{13.93}$$

The intrinsic impedance for various materials have been examined earlier. They are repeated here for reference [5]:

$$\text{Partially-conducting medium} \quad \eta = \left[\frac{j\omega\mu}{(\sigma + j\omega\varepsilon)}\right]^{1/2} \tag{13.94}$$

$$\text{Conducting medium} \quad \eta = \left(\frac{\omega\mu}{\sigma}\right)^{1/2} < 45^\circ \tag{13.95}$$

$$\text{Perfect dielectric} \quad \eta = \left(\frac{\mu}{\varepsilon}\right)^{1/2} \tag{13.96}$$

$$\text{Free space} \quad \eta_0 = \left(\frac{\mu_0}{\varepsilon_0}\right)^{1/2} = 120\pi\ \Omega \tag{13.97}$$

Standing Waves

When waves traveling in a perfect dielectric ($\sigma_1 = \alpha_1 = 0$) are normally incident on the interface with a perfect conductor ($\sigma_2 = \infty$, $\eta_2 = 0$), the reflected wave, in combination with the incident wave, produces a *standing wave.* Such a wave is readily demonstrated on clamped taut string. The oscillations at all points of half-wavelength interval are in time phase. The combination of incident and reflected waves may be written:

$$\mathbf{E}(z,t) = \left[E_0^i e^{j(\omega t-\beta z)} + E_0^r e^{j(\omega t+\beta z)}\right]\mathbf{a}_x = e^{j\omega t}\left(E_0^i e^{-j\beta z} + E_0^r e^{j\beta z}\right)\mathbf{a}_x \tag{13.98}$$

Since $\eta_2 = 0$, $E_0{}^r/E_0{}^i = -1$ and

$$\mathbf{E}(z,t) = e^{j\omega t}\left(E_0^i e^{-j\beta z} - E_0^i e^{j\beta z}\right)\mathbf{a}_x = -2jE_0^i \sin\beta z\, e^{j\omega t}\mathbf{a}_x \tag{13.99}$$

or, taking the real part,

$$\mathbf{E}(z,t) = 2E_0^I \sin\beta z \sin\omega t \, \mathbf{a}_x \tag{13.100}$$

The standing wave is shown in Figure 13.7 at time intervals of T/8, where $T = 2\pi/\omega$ or $1/f$ is the period. At $t = 0$, $\mathbf{E} = 0$ everywhere; at $t = 1(T/8)$, the end points of the **E** vectors lie on sine curve 1; at $t = 2(T/8)$, they lie on curve 2; and so forth. Sine curves 2 and 6 form an envelope for the oscillations, the amplitude of this envelope is twice the amplitude of the incident wave. Note that adjacent half-wave-length segments are 180° out of phase with each other [5].

Power and the Poynting Vectors

Maxwell's first equation for a region with conductivity σ is written and then **E** is dotted with each term:

$$\nabla \times \mathbf{H} = \sigma\mathbf{E} + \varepsilon\left(\frac{\partial \mathbf{E}}{\partial t}\right) \tag{13.101}$$

$$\mathbf{E} \cdot (\nabla \times H) = \sigma E^2 + \mathbf{E} + \varepsilon\left(\frac{\partial \mathbf{E}}{\partial t}\right) \tag{13.102}$$

where, as usual,

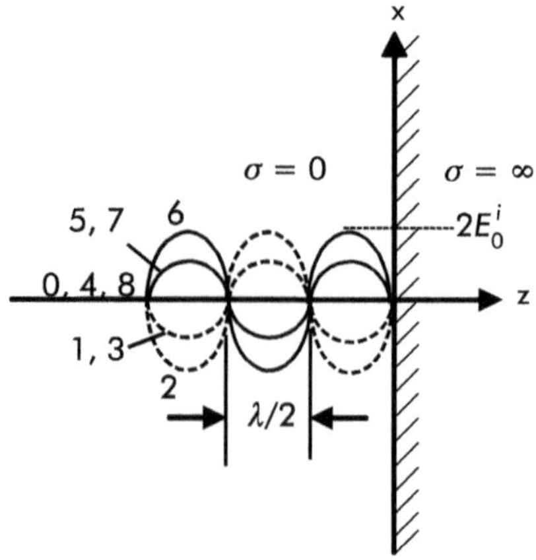

Figure 13.7 Illustration of a standing wave.

$$E^2 = \mathbf{E} \cdot \mathbf{E} \tag{13.103}$$

The vector identity:

$$\nabla \cdot (\mathbf{A} \times \mathbf{B}) = \mathbf{B} \cdot (\nabla \times \mathbf{A}) - \mathbf{A} \cdot (\nabla \times \mathbf{B}) \tag{13.104}$$

is employed to change the left side of the equation,

$$\mathbf{H} \cdot (\nabla \times \mathbf{E}) - \nabla \cdot (\mathbf{E} \times \mathbf{H}) = \sigma E^2 + \mathbf{E} + \varepsilon\left(\frac{\partial \mathbf{E}}{\partial t}\right) \tag{13.105}$$

By Maxwell's second equation,

$$\mathbf{H} \cdot (\nabla \times \mathbf{E}) = \mathbf{H} \cdot \left[-\mu\left(\frac{\partial \mathbf{E}}{\partial t}\right)\right] = -\frac{1}{2}\mu\left(\frac{\partial H^2}{\partial t}\right) \tag{13.106}$$

Similarly,

$$\mathbf{E} \cdot \varepsilon\left(\frac{\partial \mathbf{E}}{\partial t}\right) = \frac{1}{2}\varepsilon\left(\frac{\partial E^2}{\partial t}\right) \tag{13.107}$$

Substituting and rearranging terms,

$$\sigma E^2 = -\frac{1}{2}\varepsilon\left(\frac{\partial E^2}{\partial t}\right) - \frac{1}{2}\mu\left(\frac{\partial H^2}{\partial t}\right) - \nabla \cdot (\mathbf{E} \times \mathbf{H}) \tag{13.108}$$

If this equality is valid, then the integration of its terms throughout a general volume V must also be valid.

$$\int_V \sigma E^2\, dv = -\int_V \left[\frac{1}{2}\varepsilon\left(\frac{\partial E^2}{\partial t}\right) + \frac{1}{2}\mu\left(\frac{\partial H^2}{\partial t}\right)\right] dv - \oint_S\!\!\int (\mathbf{E} \times \mathbf{H}) \cdot d\mathbf{S} \tag{13.109}$$

where the last term has been converted to an integral over the surface of v by use of the divergence theorem [5].

The integral on the left has the units of watts and is the usual ohmic term representing energy dissipated per unit time in heat.

This dissipated energy has its source in the integrals on the right. Because $\frac{1}{2}\varepsilon E^2$ and $\frac{1}{2}\mu H^2$ are the densities of energy stored in the electric and magnetic fields, respectively, the negative time derivatives can be viewed as a decrease in this stored energy. Consequently, the final integral (including the minus sign) must be the rate of energy entering the volume from the outside. A change of sign produces the *instantaneous rate of energy leaving the volume*:

$$P(t) = \oint_s (\mathbf{E} \times \mathbf{H}) \cdot d\mathbf{S} = \oint_s \mathbf{P} \cdot d\mathbf{S} \tag{13.110}$$

where:

$$\mathbf{P} = \mathbf{E} \times \mathbf{H} \tag{13.111}$$

is the *Poynting vector*. It is the instantaneous rate of energy flow per unit area at a point distant from the originating point.

In the cross product that defines the Poynting vector, the fields are supposed to be in real form. If, instead, **E** and **H** are expressed in complex form and have common time-dependence, $e^{j\omega t}$, then the time-average of **P** is given by:

$$\mathbf{P}_{avg} = \frac{1}{2}\mathrm{Re}\left(\mathbf{E} \times \mathbf{H}^*\right) \tag{13.112}$$

where $\mathbf{H}^*$ is the complex conjugate of **H**. This follows the *complex power* of circuit analysis,

$$\mathbf{S} = \frac{1}{2}\mathbf{VI}^* \tag{13.113}$$

of which the power is the real part,

$$\mathbf{P}_{avg} = \frac{1}{2}\mathrm{Re}\,\mathbf{VI}^* \tag{13.114}$$

For plane waves, the direction of energy flow is the direction of propagation. Thus, the Poynting vector offers a useful, coordinate-free way of specifying the direction of propagation or of determining the directions of the fields if the direction of propagation is known. This can be particularly valuable where incident, transmitted, and reflected waves are being examined.

References

[1] Shadowitz, A., *The Electromagnetic Field*, New York, Dover Publications, Inc., 1975.

[2] Jordan, E. C., *Electromagnetic Waves and Radiating Systems*, London, Constable and Company, Ltd., 1953.

[3] Russer, P., *Electromagnetics, Microwave Circuit and Antenna Design for Communications Engineering*, Norwood, MA: Artech House, 2006.

[4] Warnick, K. F., and P. Russer, *Problem Solving in Electromagnetics, Microwave Circuit and Antenna Design for Communications Engineering*, Norwood, MA: Artech House, 2006.

[5] Cheng, D. K., *Fundamentals of Engineering Electromagnetics*, Reading, MA: Addison-Wesley Publishing Company, Inc., 1993.

[6] Macnamara, T., *Handbook of Antennas for EMC*, Norwood, MA: Artech House, 1995.

[7] Friesen, D. J., *Introduction to Antennas and RF Propagation Analysis*, Norwood, MA: Artech House, 2022.

[8] Blake, L. V., *Antennas*, Norwood, MA: Artech House, 1984.

14

BOUNDARY CONDITIONS

14.1 INTRODUCTION

The behavior of the electric field intensity **E** and the electric flux density **D** across the interface of two different materials was examined earlier, where in Chapter 7, static fields were the topic of discussion. Likewise, a similar treatment will again be given for static fields. However, this time the focus is on the magnetic field strength **H** and the magnetic flux density **B**. This will then complete the study of the boundary conditions on the four principal vector fields; that is, **E**, **D**, **H**, and **B** [5]. Units for these vectors are: **E** (V/m), **D** (C/m^2), **H** (A/m), and **B** (Tesla or T).

In Chapter 13, time variable fields were treated, displacement current density $\mathbf{J}_D$ was introduced, and Faraday's Law was examined. In this chapter, these same equations, and others developed earlier, are grouped together to form the equation set known as *Maxwell's Equations* [5]. These equations underlie the whole of electromagnetic field theory.

14.2 BOUNDARY RELATIONS FOR MAGNETIC FIELDS

When **H** and **B** are examined at the interface between two different materials, abrupt changes can be expected. The changes are similar to those noted for **E** and **D** at the interface between the two different dielectrics [1].

In Figure 14.1 an interface is shown separating Material #1, having properties σ_1 and μ_{r1}, from Material #2 with its properties σ_2 and μ_{r2}. The behavior of **B** can be determined by use of a small right circular cylinder positioned across the interface as shown in the figure. Since magnetic lines of flux are continuous,

$$\oint \mathbf{B} \cdot d\mathbf{S} = \int \mathbf{B}_1 \cdot d\mathbf{S}_1 + \int \mathbf{B}_{\text{cylinder}} \cdot d\mathbf{S}_{\text{cylinder}} + \int \mathbf{B}_2 \cdot d\mathbf{S}_2 = 0 \tag{14.1}$$

If the two planes are allowed to approach one another and keeping the interface between them, the area of the curved surface will approach zero, giving

$$\int \mathbf{B}_1 \cdot d\mathbf{S}_1 + \int \mathbf{B}_2 \cdot d\mathbf{S}_2 = 0 \tag{14.2}$$

or

$$-B_{n1} \int dS_1 + B_{n2} \int dS_2 = 0 \tag{14.3}$$

from which

$$B_{n1} = B_{n2} \tag{14.4}$$

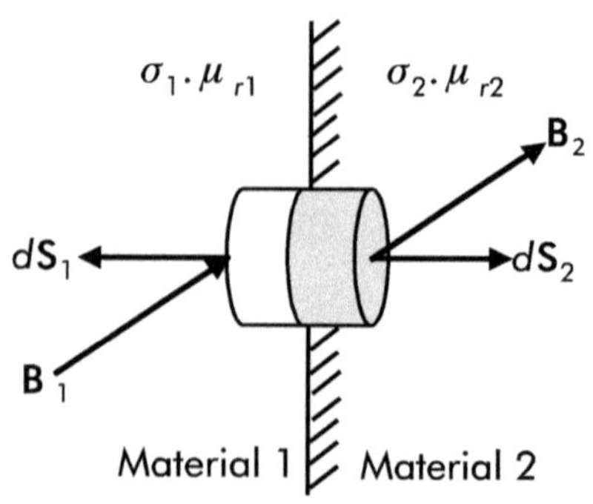

Figure 14.1 Illustration of magnetic field boundary relations.

In words, the *normal component* of **B** is continuous across an interface. Note that either normal to the interface may be used in calculating B_{n1} and B_{n2} [2].

The variation in **H** across an interface is obtained by the application of Ampere's Law around a closed rectangular path as shown in Figure 14.2.

Assuming no current at the interface and letting the rectangular shrink to zero in the usual way,

$$0 = \oint \mathbf{H} \cdot d\mathbf{l} \rightarrow H_{l1}\Delta l_1 - H_{l2}\Delta l_2 \tag{14.5}$$

where

$$H_{l1} = H_{l2} \tag{14.6}$$

Thus, tangential **H** has the same projection along the two sides of the rectangle. Since the rectangle can be rotated 90° and the argument repeated, it follows that:

$$H_{t1} = H_{t2} \tag{14.7}$$

In words, the *tangential component* of **H** is continuous across a current-free interface.

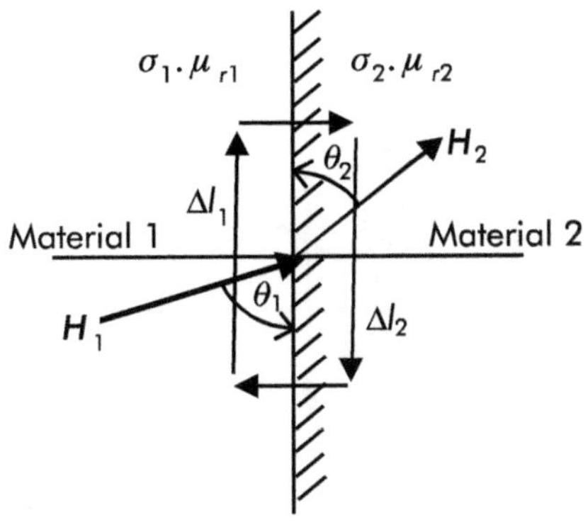

Figure 14.2 Ampere's Law applied around a closed rectangular path.

The relation

$$\frac{\tan\theta_1}{\tan\theta_2} = \frac{\mu_{r2}}{\mu_{r1}} \tag{14.8}$$

between the angles made by $\mathbf{H}_1$ and $\mathbf{H}_2$ with a current-free interface (see Figure 14.2) is obtained [4].

14.3 CURRENT SHEET AT THE BOUNDARY

If one material at the interface has a nonzero conductivity, a current may be present [5]. This could be a current throughout the material. However, of greater interest is the case of a current sheet at the interface (see Figure 14.3). Figure 14.3 shows a uniform current sheet. In the indicated coordinated system, the current sheet has density

$$\mathbf{K} = K_0\mathbf{a}_y \tag{14.9}$$

and it is located at the interface $x = 0$ between regions 1 and 2. The magnetic field $\mathbf{H}'$ produced by this current sheet is given by

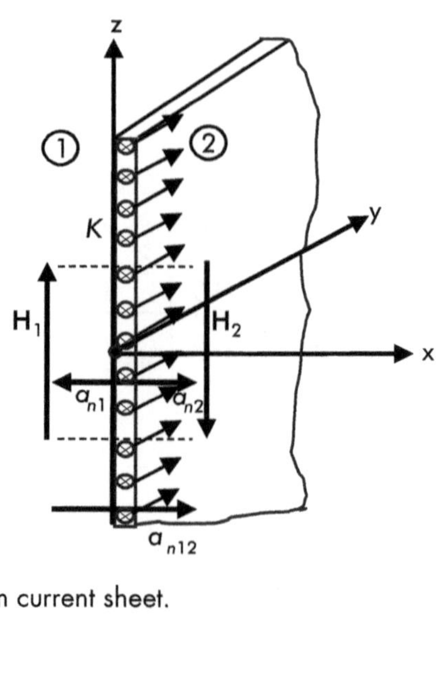

Figure 14.3 Uniform current sheet.

$$\mathbf{H}_1' = \frac{1}{2}\mathbf{K} \times \mathbf{a}_{n1} = \frac{1}{2}K_0\mathbf{a}_z \tag{14.10}$$

and

$$\mathbf{H}_2' = \frac{1}{2}\mathbf{K} \times \mathbf{a}_{n2} = \frac{1}{2}K_0\left(-\mathbf{a}_z\right) \tag{14.11}$$

Thus, $\mathbf{H}'$ has a tangential discontinuity of magnitude $|K_0|$ at the interface. If a second magnetic field, $\mathbf{H}''$, arising from some other source, is present, its tangential component will be continuous at the interface. The resultant magnetic field,

$$\mathbf{H} = \mathbf{H}' + \mathbf{H}'' \tag{14.12}$$

will then have a discontinuity of magnitude $|K_0|$ in its tangential component. (Note: The unit for K_0 is A/m.) This is expressed by the vector formula:

$$\left(\mathbf{H}_1 - \mathbf{H}_2\right) \times \mathbf{a}_{n12} = \mathbf{K} \tag{14.13}$$

where $\mathbf{a}_{n12}$ is the unit normal from region 1 to region 2. The vector relation, which is independent of the choice of coordinate system, also holds for a nonuniform current sheet, where $\mathbf{K}$ is the value of the current density at the considered interface point.

14.4 SUMMARY OF BOUNDARY CONDITIONS

For reference purposes, the relationships for **E** and **D** (electric fields) across the interface of two dielectrics are shown below along with the relationships for **H** and **B** [6].

Magnetic Fields

$$B_{n1} = B_{n2} \tag{14.14}$$

$$H_{t1} = H_{t2} \quad \text{(current free)} \tag{14.15}$$

$$(\mathbf{H}_1 - \mathbf{H}_2) \times \mathbf{a}_{n12} = \mathbf{K} \quad \text{(with a current sheet)} \tag{14.16}$$

$$\frac{\tan\theta_1}{\tan\theta_2} = \frac{\mu_{r2}}{\mu_{r1}} \quad \text{(current free)} \tag{14.17}$$

Electric Fields

$$D_{n1} = D_{n2} \quad \text{(charge free)} \tag{14.18}$$

$$(\mathbf{D}_1 - \mathbf{D}_2) \cdot \mathbf{a}_{n12} = -\rho_s \quad \text{(with surface charge)} \tag{14.19}$$

$$E_{t1} = E_{t2} \tag{14.20}$$

$$\frac{\tan\theta_1}{\tan\theta_2} = \frac{\varepsilon_{r2}}{\varepsilon_{r1}} \quad \text{(charge)} \tag{14.21}$$

These relationships were obtained under the assumption of static conditions. However, they also apply to time-variable fields as well.

References

[1] Shadowitz, A., *The Electromagnetic Field*, New York, Dover Publications, Inc., 1975.

[2] Jordan, E. C., *Electromagnetic Waves And Radiating Systems*, London, Constable and Company, Ltd., 1953.

[3] Russer, P., *Electromagnetics, Microwave Circuit and Antenna Design for Communications Engineering*, Norwood, MA: Artech House, 2006.

[4] Warnick, K. F., Russer, P., *Problem Solving in Electromagnetics, Microwave Circuit and Antenna Design for Communications Engineering*, Norwood, MA: Artech House, 2006.

[5] Cheng, D. K., *Fundamentals of Engineering Electromagnetics*, Reading, MA: Addison-Wesley Publishing Company, Inc., 1993.

[6] Macnamara, T., *Handbook of Antennas for EMC*, Norwood, MA: Artech House, 1995.

15

MAXWELL'S EQUATIONS

Maxwell's equations are based on the individual efforts of Carl Frederic Gauss, Michael Faraday, Jean-Baptiste Biot, and Felix Savart. As such, these efforts were made well prior to the collective work of James Clerk Maxwell. The general form of Maxwell's Equations, when grouped into a general set, can be displayed as shown in Tables 15.1 and 15.2 [4].

Table 15.1
Maxwell's Equations: General Set

Point Form	Integral Form		
$\nabla \times \bar{\mathbf{H}} = \bar{\mathbf{J}}_c + \frac{\partial \bar{\mathbf{D}}}{dt}$	$\oint \bar{\mathbf{H}} \cdot d\mathbf{l} = \int_S \left(\mathbf{J}_c + \frac{\partial \bar{\mathbf{D}}}{\partial t} \right) \cdot d\bar{\mathbf{S}}$	(Ampere's law)	(15.1)
$\nabla \times \bar{\mathbf{E}} = -\frac{\partial \bar{\mathbf{B}}}{dt}$	$\oint \bar{\mathbf{E}} \cdot d\mathbf{l} = \int_S \left(-\frac{\partial \bar{\mathbf{B}}}{\partial t} \right) \cdot d\bar{\mathbf{S}}$	(Faraday's law; S fixed)	(15.2)
$\nabla \cdot \bar{\mathbf{D}} = \rho$	$\oint_S \bar{\mathbf{D}} \cdot d\bar{\mathbf{S}} = \int_s \rho \, dv$	(Gauss's law)	(15.3)
$\nabla \cdot \bar{\mathbf{B}} = 0$	$\oint_S \bar{\mathbf{B}} \cdot d\bar{\mathbf{S}} = 0$	(Nonexistence of magnetic monopole)	(15.4)

For free space conditions where there are no charges ($\rho = 0$) and no conduction currents ($\mathbf{J}_c = 0$), Maxwell's equations take the form of the free-space set shown next [4].

Table 15.2
Maxwell's Equations: Free Space Set

Point Form	Integral Form	
$\nabla \times \bar{\mathbf{H}} = \dfrac{\partial \bar{\mathbf{D}}}{dt}$	$\oint \bar{\mathbf{H}} \cdot d\mathbf{l} = \int_S \left(\dfrac{\partial \bar{\mathbf{D}}}{\partial t} \right) \cdot d\bar{\mathbf{S}}$	(15.5)
$\nabla \times \bar{\mathbf{E}} = -\dfrac{\partial \bar{\mathbf{B}}}{dt}$	$\oint \bar{\mathbf{E}} \cdot d\mathbf{l} = \int_S \left(-\dfrac{\partial \bar{\mathbf{B}}}{\partial t} \right) \cdot d\bar{\mathbf{S}}$	(15.6)
$\nabla \cdot \bar{\mathbf{D}} = 0$	$\oint_S \bar{\mathbf{D}} \cdot d\bar{\mathbf{S}} = 0$	(15.7)
$\nabla \cdot \bar{\mathbf{B}} = 0$	$\oint_S \bar{\mathbf{B}} \cdot d\bar{\mathbf{S}} = 0$	(15.8)

Discussions for each of the four basic Maxwell equations are given as shown in Sections 15.1 to 15.4.

15.1 MAXWELL'S FIRST EQUATION

Maxwell's first equation is based on Ampere's law [3].

Ampere's law states that the line integral of the magnetic field intensity is equal to the total enclosed current (i.e., the sum of the conduction and displacement currents). It is expressed by:

$$\oint \mathbf{H} \cdot d\mathbf{l} = I_{\text{conduction}} + I_{\text{displacement}} \tag{15.9}$$

$\mathbf{H}$ = magnetic field intensity vector (A/m)

$d\mathbf{l}$ = elementary length vector (m)

$I_{\text{conduction}}$ = conduction current (A)

$I_{\text{displacement}}$ = displacement current (A)

The equation, when expressed in terms of conduction and displacement currents, can also be written in terms of the area and the conduction and displacement current densities, $\mathbf{J}_c$ and $\mathbf{J}_d$, respectively:

$$\begin{aligned}\oint \mathbf{H} \cdot d\mathbf{l} &= \int_S \left(\mathbf{J}_c + \mathbf{J}_d\right) dS \\ &= \sigma \mathbf{E} \\ &= \frac{\partial \mathbf{D}}{\partial t} \\ &= \frac{\partial}{\partial t}(\varepsilon \mathbf{E})\end{aligned} \tag{15.10}$$

$\mathbf{H}$ = magnetic field intensity vector (A/m)

$d\mathbf{l}$ = elementary length vector (m)

$\mathbf{J}_c$ = conduction current density vector (A/m^2).

σ = conductivity (S/m)

$\mathbf{E}$ = electric field intensity (V/m)

$\mathbf{J}_d$ = displacement current density vector (A/m^2)

$\partial \mathbf{D}/\partial t$ = partial derivative of electric flux density, $\mathbf{D}$, with respect to time (C/m^2)

ε = permittivity (F/m)

$\mathbf{D}$ = density of electric flux (C/m^2)

<= 1st partial derivative with respect to time

ε = permittivity (F/m)

$\mathbf{E}$ = electric field intensity (V/m)

dS = element of area perpendicular to the current flow (m^2)

$$= \int \left[\sigma \mathbf{E} + \frac{\partial}{\partial t}(\varepsilon \mathbf{E})\right] dS \quad \leftarrow \text{(Integral form, Maxwell's first equation)} \tag{15.11}$$

The integral form, shown in the text above, can be written in words as: "The magnetomotive force around a closed path is equal to the conduction current plus the time derivative of the electric flux density through any surface bounded by the path" [2].

Applying Stoke's theorem next, we get:

$$\int_S \text{curl } \mathbf{H} \cdot d\mathbf{S} = \int_S \left[\sigma \mathbf{E} + \frac{\partial}{\partial t}(\varepsilon \mathbf{E})\right] dS \tag{15.12}$$

$$\text{curl}\,\mathbf{H} = \sigma\mathbf{E} + \varepsilon\frac{\partial}{\partial t}(\mathbf{E})\text{, that is, the differential vector form} \tag{15.13}$$

of Maxwell's first equation.

Continuing, Maxwell's first equation, written in scalar differential form for each component of the magnetic field intensity (H) using the definition of the component of the curl, takes on the following form:

$$\frac{\partial}{\partial y}(H_z) - \frac{\partial}{\partial z}(H_y) = \varepsilon\frac{\partial}{\partial t}(E_x) + \sigma E_x \tag{15.14}$$

$$\frac{\partial}{\partial z}(H_x) - \frac{\partial}{\partial x}(H_y) = \varepsilon\frac{\partial}{\partial t}(E_y) + \sigma E_x \tag{15.15}$$

$$\frac{\partial}{\partial y}(H_y) - \frac{\partial}{\partial x}(H_x) = \varepsilon\frac{\partial}{\partial t}(E_z) + \sigma E_z \tag{15.16}$$

For further discussion on the subject of Gauss' law, the reader is encouraged to review Chapter 3.

15.2 MAXWELL'S SECOND EQUATION

The second equation is based on Faraday's law.

Faraday's law states "a magnetic flux of Φ induces an electromotive force (EMF) in a conductor" [1].

This induced EMF has a magnitude V that is proportional to the rate of change of the magnetic flux Φ as given by:

$$V = -\frac{\partial}{\partial t}(\Phi) \tag{15.17}$$

The negative sign indicates that the voltage opposes the flux that is causing it. The voltage is also equal to the line integral of the electric field $\mathbf{E}$ over the length of the conductor:

$$|V| = \int \mathbf{E} \cdot d\mathbf{l} \tag{15.18}$$

$\mathbf{E}$ = electric field intensity (V/m)

$d\mathbf{l}$ = elementary vector length (m) of the total loop

Thus, we can say that

$$\oint \mathbf{E} \cdot d\mathbf{l} = -\frac{\partial}{\partial t}(\Phi) \tag{15.19}$$

$$\Phi = \int \mathbf{B} \cdot d\mathbf{S} \tag{15.20}$$

$\mathbf{B}$ = magnetic flux density (Tesla)

$d\mathbf{S}$ = the vector area normal to the flux density $\mathbf{B}$

Combining the two equations, we get Maxwell's second equation in integral form:

$$\oint_{\text{perimeter}} \mathbf{E} \cdot d\mathbf{l} = -\int_{\text{surface}} \frac{\partial}{\partial t}(\mathbf{B} \cdot d\mathbf{S}) \tag{15.21}$$

The above expression states: "The EMF around a closed path is equal to the time derivative of the magnetic displacement through any surface bounded by the path."

By analogy with electric current, we can call the time derivative of the magnetic flux, the magnetic current. Thus, we can also state Maxwell's second equation as: "The electric voltage around a closed path is equal to the magnetic current through the path."

Maxwell's equation in differential form is derived by way of using Stoke's theorem which states: "The curl of a vector is the maximum value of the line integral around an elementary surface as the surface shrinks to zero."

Thus, the curl is given by:

$$\text{curl } \mathbf{A} = \text{limit} \frac{\oint A \cdot d\mathbf{l}}{\delta S} \tag{15.22}$$

$$\oint_{\text{perimeter}} \mathbf{E} \cdot d\mathbf{l} = -\int_{\text{surface}} \text{curl } \mathbf{E} \cdot d\mathbf{S} \tag{15.23}$$

Maxwell's second equation in differential form is expressed as:

$$\text{curl}\,\mathbf{E} = -\frac{\partial}{\partial t}(\mathbf{B}) \tag{15.24}$$

$\mathbf{E}$ = electric field intensity (V/m)

$\mathbf{B}$ = magnetic flux density (Tesla)

15.3 MAXWELL'S THIRD EQUATION

Maxwell's third equation is based on Gauss' theorem.

Gauss' theorem states: "The total outward flux is proportional to the total charge enclosed."

In the case of Gauss' theorem, the divergence of the electric flux density, $\mathbf{D}$, must be considered [1]. Writing out the total outward flux in terms of this divergence we have

$$\text{total flux} = \text{div}\,\mathbf{D}\int_{V_u} dV_u = \int_{V_u} \rho\, dV_u \tag{15.25}$$

$\mathbf{D}$ = electric flux density (C/m^2)

ρ = electric flux density (C/m^2)

dV_u = volume of infinitesimal size (m^3)

If one considers a volume that has an infinitesimal size of ΔV_u, it can be assumed that the charge distribution ρ is constant over it, and (15.25) can be written as:

$$\text{div}\,\mathbf{D}\int_{\Delta V_u} dV_u = \rho\int dV_u \tag{15.26}$$

which reduces to Maxwell's third equation which is, namely:

$$\text{div}\,\mathbf{D} = \rho \tag{15.27}$$

In integral form, this equation can be expressed as:

$$\int_S \mathbf{D} \cdot d\mathbf{S} = \int_{V_u} \rho \, dV_u \tag{15.28}$$

$d\mathbf{S}$ = vector area normal to $\mathbf{D}$

Note: $\mathbf{D}$, dV_u, and ρ are defined in the text of (15.25).

S is a closed surface. Maxwell's third equation can be stated as follows: "The total electric displacement through the surface enclosing a volume is equal to the total charge within the volume" [2].

15.4 MAXWELL'S FOURTH EQUATION

Maxwell's fourth equation is the magnetic flux equivalent of the electric flux as defined by the third equation. Divergence of the magnetic flux density **B** is equal to the distribution of the magnetic poles enclosed [3]. The fourth equation states: "The divergence of magnetic flux density **B** is zero."

This is not unexpected since there are no free north or south magnetic poles. The flux from the north pole travels to the south pole of a magnet, and a south pole travels to the north pole of a magnet. As such the poles always appear is pairs, making the resultant magnetic pole strength zero. There are no magnetic monopoles in the electromagnetic physical world.

In differential vector form, Maxwell's fourth equation can be written as:

$$\text{div}\, \mathbf{B} = 0 \tag{15.29}$$

$\mathbf{B}$ = magnetic flux density (Tesla)

In integral form, the fourth equation can also be expressed as:

$$\int \mathbf{B} \cdot d\mathbf{S} = 0 \tag{15.30}$$

$\mathbf{B}$ = magnetic flux density (Tesla)

$d\mathbf{S}$ = the elemental vector of the closed surface area normal to $\mathbf{B}$

It can finally be stated for Maxwell's fourth equation that net magnetic flux emerging through any closed surface is 0.

15.5 OTHER IMPORTANT EQUATIONS

There are two electromagnetically vital equations that are left to be shared [2]:

$$\text{Fifth equation: } \mathbf{D} = \varepsilon_o \mathbf{E} \tag{15.31}$$

$\mathbf{D}$ = electric flux density (C/m^2)

ε_o = electric permittivity of free space = 8.854×10^{-12} (F/m)

$\mathbf{E}$ = electric field (V/m)

$$\text{Sixth Equation: } \mathbf{B} = \mu_o \mathbf{H} \tag{15.32}$$

$\mathbf{B}$ = magnetic flux density (Tesla)

μ_o = magnetic permeability of free space = $4\pi \times 10^{-7}$ (H/m)

$\mathbf{H}$ = magnetic field (A/m)

15.6 A SUMMARY OF THE PHYSICS-BASED RAMIFICATIONS OF MAXWELL'S EQUATIONS

1. A static **E** field can exist in the absence of a magnetic field **H.** A capacitor with a static charge *Q* furnishes an example.
2. Likewise, a conductor with a constant current, *I*, has a magnetic field **H** without an **E** field.

   When fields are time-variable, however, **H** cannot exist without an **E** field, nor can **E** exist without a corresponding **H** field.
3. While much valuable information can be derived from static field theory, only with time-variable fields can the full value of electromagnetic field theory be demonstrated.
4. The experiments of Faraday and Hertz and the theoretical analyses of Maxwell all involve time-variable fields.

5. There are no magnetic monopoles, only magnetic dipoles, in EM physics.

References

[1] Shadowitz, A., *The Electromagnetic Field*, New York, Dover Publications, Inc., 1975.

[2] Jordan, E. C., *Electromagnetic Waves and Radiating Systems*, London, Constable and Company, Ltd., 1953.

[3] Cheng, D. K., *Fundamentals of Engineering Electromagnetics*, Reading, MA: Addison-Wesley Publishing Company, Inc., 1993.

[4] Macnamara, T., *Handbook of Antennas for EMC*, Norwood, MA: Artech House, 1995.

16

PRACTICAL SOLUTIONS TO REAL-WORLD ELECTROMAGNETIC ENGINEERING PROBLEMS

INTRODUCTION

My own past experience with books that address the subject of EM principles and theory is that none ever included a set of anecdotes that described practical solutions to real-world workplace electromagnetic problems. This is most unfortunate. In engineering school, one of my most memorable mechanical engineering classes was taught by a 25-year Pennsylvania steels engineer whose oral workplace anecdotes made what would have been an otherwise boring subject (to me, anyway) truly more enjoyable to learn. The subject of electromagnetics is for many engineering students a required course that is truly dreaded.

I believe that most books on EM theory are written by academicians and not workplace-experienced engineers/physicists. The problem is that few academicians have practical problem-solving experience from which to create meaningful anecdotes. On the other hand, from my own experience the industrial workplace day used to get so full just doing the basic engineering work and solving (and also helping others solve) EM problems that there was little or no inclination to publish a book. Once I retired, I decided to try to make a difference.

Thus, one of the key reasons why I wrote this two-edition book set was to give the workplace engineer an easy-to-read, quick, handy reference set that contained both EM theory and included a discussion of real-world EM engineering problem anecdotes. In my view, real-world workplace EM-related problems are among the most difficult engineering problems to solve. Accordingly, these anecdotes were written and published in the hope that they would provide the engineering professional greater insight into how to identify, isolate, and solve EM-related problems at work. These anecdotes I offer came from my true-to-life situations.

As such, this purely anecdotal section is divided into three subject problem areas:

1. Electrostatics;
2. Magnetostatics;
3. Electromagnetics.

I wish to convey to the reader my sincere hope that the contents of this section are truly helpful to the workplace engineer or physicist. Be rest assured that there are plenty of workplace situations that can truly baffle even the most clever, experienced practicing engineer. Truly, no one is immune.

The first piece of advice I give to the engineer facing a truly baffling engineering problem and is being pressed by management to quickly solve the problem is: Work to stay calm.

As hard as it sometimes is, staying calm will nearly always help to keep your head clear. Furthermore, it will help keep your evenings and sleeping nights peaceful. I say this knowing full well that sometimes this can be more easily said than done.

I have found across my 37 professional years that my finest, swiftest troubleshooting/problem-solving abilities came about when I made the deliberate, concerted effort to keep cool even while management pressed me for answers. It may also be—and I hope it is not your case—that you will encounter highly volatile management types in your career (i.e., leads, supervisors, or managers). A poem entitled "If" written by Rudyard Kipling written many years ago contains two lines that have become my favorite:

> *If you can keep your head when all about you are losing theirs and blaming it on you. . . you'll be a Man, my son!*
>
> —Rudyard Kipling (1865–1935)

I suggest putting these lines on a piece of office decoration and placing them in plain view to you in your workstation, office cube, or home office desk. View it as often as you find the need to do so. Others visiting you may read it and find inspiration, too.

16.1 ANECDOTES IN ELECTROSTATICS

The path leading to my introduction to the field of commercial jet airplane fuel tank electrostatics was neither desired nor welcome. I used to think to myself as I studied electrostatics, "When will I ever use *this* in the workplace?"

As the newest member of the Electromagnetic Effects (EME) and Antennas group, Product Development division (R&D) of Boeing Commercial Airplanes in 2016, I was about to find out.

In 2016, I had just completed a successful four-year long stint as the lead engineer for the 737MAX Electromagnetic Effects (EME) Team at Boeing Commercial Airplanes (BCA). In June of that year my manager, Stan, asked if I wanted to become a Level 1 manager at Boeing. While highly complimented for being asked, I had absolutely no interest at all in going into engineering management. My true engineering love was solving problems and performing RF/EM-related testing.

I wanted to return to the technical side of engineering. Solving problems, even EME-related problems, was far more enjoyable to me than the thought of managing people. Besides, prior to coming to BCA (in 2012), I had been doing technical work for Boeing Defense Systems (BDS) and other US defense engineering companies for 25 years, and I wanted to resume what I did well and greatly enjoyed.

When I found myself moving over to a BCA product development (PD) organization and, specifically, to the engineering team called the EME and Antennas Group, you can't imagine how overjoyed I was. Moreover, the Harbour Pointe building at which I arrived to work in Mukilteo, WA (Seattle/Puget Sound area) faced West and overlooked the absolutely gorgeous Puget Sound waters. Through the large west-facing clear-glass picture windows of our building, I could see sailboats, large freight barges, and other recreational pleasure craft passing by at lunchtime. I used to sit and look out this window for a half hour every day. It helped me relax, keeping me more serene during the workday.

However, my new boss, Mark, told me on the very first day that he was giving me his team's hardest technical problems.

"Great," I thought, "nothing like throwing me into the fire on my first day."

I was to find that this was typical of my new boss' leadership style, and in time I found that he would throw many technical challenges my way. In looking back, I came to realize that his doing this was the highest compliment he could pay an engineer.

The first R&D task he assigned to me was an electrostatics-based project, one in which the soon-to-be-retiring PhD research engineer simply was not able to solve for over three years. The overarching goal of this first R&D project was to figure out how to reduce 787 airplane fuel tank production costs by eliminating a particular airplane tank manufacturing process that was costing BCA greatly (intellectual property security concerns prevent me from disclosing the details). Suffice it to say that the process was quite effective but tremendously expensive to manufacture. Cost concerns drove management to seek a cost-reduction solution.

Immediately, I could see that I had a lot of reading and research to do on composite fuel tanks (i.e., their design, production, and airplane installation). Serendipitously, in earlier years from 2004 to 2011, I completed a thorough personal study on electromagnetics and vector calculus. I grew to discover that I had mastered electrostatics, magnetostatics, and EM waves quite well, and I was so glad for that. So, having reference books, a good memory, and notes in three-ring binders to draw from, I discovered that I was better-armed to take on this problem than I knew at the start. Here is a brief description on the pending electrostatics problem.

16.1.1 High Electrostatic Charge Removal from Airplane Fuels Tanks After Post Fuel (Jet A) Refueling

16.1.1.1 Problem

The initial R&D problem put before me was to figure out a more effective way to either prevent or remove the high amounts of electrostatic charge that was entering the composite (CFRP) material airplane fuel tank as each airplane was being refueled between flights. The fear from the FAA was that electrostatic charge in the vaporous ullage of an airplane Jet A fuel tank had the potential to blow the airplane up in-flight. An added complication was a new FAA regulation that was at the time being added to the roster of new FAA CFRs in 2016. The new regulation said, in essence, that the probability of airplane fuel vapor ignition in airplane fuel tank ullages had to be zero. I thought, "You have *got* to be kidding!"

I knew instinctively, as did several of my PhD physicist colleagues, that there is *no* way to make the probability of fuel tank ullage ignition zero. It reminded me of the engineer joke that said that, in order to solve certain engineering problems, it is first required to procure an *extremely* rare material called "unobtain-ium." Well, this outrageous electrostatics-based FAA regulation for airplane fuel tanks was truly a classic example of this joke brought to life.

Simply put, there is no way to cost-effectively quantify/verify that the probability of in-tank airplane tank ullage ignition is zero. You can aim to make it close to zero, but it will never be zero. Zero is unobtainable. However, because the FAA did in fact have this requirement at the time (it was revised five years later), we were legally bound by US law to seek a solution to the "problem."

So, now let us "talk through" the problem.

16.1.1.2 Background

Liquid dielectric Jet A fuel (i.e., kerosene with various conductive fuel additives) passes through an airport tarmac refueling hose to enter the airplane fuel tank through its refueling entry port. As fuel enters the tank, it must pass through in-line particulate filters. These filters remove unwanted debris that might enter the fuel tank interior (and potentially clog fuel lines leading to the engine). Unfortunately, the process of debris removal includes the unavoidable consequence of triboelectric effect occurring in the filters as the liquid dielectric Jet A fuel passes through them. The result is that large amounts of positive electrostatic charge are created in the fuel filter, are moved down the refueling hose (with the airplane fuel), and enter the airplane fuel tank (see Figure 16.1).

In accordance with the circa 2016 version of federal regulation 14 CFR 25.981, high electric charge cannot be present in an airplane fuel tank while the airplane is either in flight or on the ground. As such, the challenge for fuel systems designers was to figure out how to either keep electric charge from entering the tank or somehow remove it from the tank after entry.

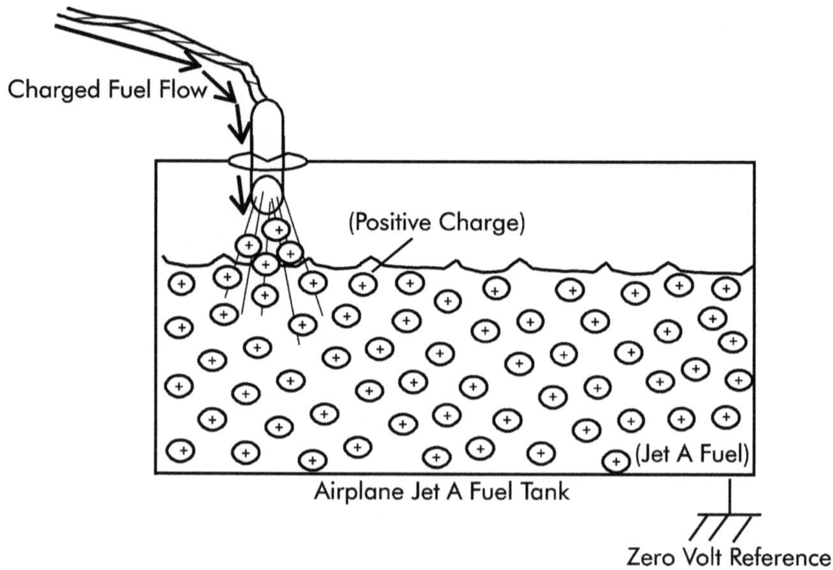

Figure 16.1 Illustration of high electrostatic charge and Jet A fuel entering a commercial airplane fuel tank.

16.1.1.3 Solutions Investigated

One option that was explored involved figuring out a way to prevent charge from entering the fuel tank in the first place. It was postulated by a previous principal investigator that a conductive charge bleed-off line could be electrically tied to a metal braid shield situated on the inside of the fuel hose. The opposite end of the line electrically would be tied to the tarmac earth ground. By doing that, the high charge could then be bled off (or "bled away") before it entered the fuel tank.

The conductive inner hose wall could then be attached to the airport earth ground by way of a conductive wire harness connected in series with high value resistors (example, 1 MΩ). This option (i.e., the preventative charge accumulation option) was thought to be a good one for US airports if introduced. However, it was mentioned that there was no way one could ensure that airport refueling stations off the continental United States (i.e., other countries and continents) would employ the potentially higher-priced refueling hoses.

Also, when you think about it, what would guarantee that the fuel hose conductive charge bleed-away line would always be attached by foreign airport ground crews to earth ground tarmac points at every airport refueling facility in the world? Nothing. Simply put, there is no guarantee. You cannot impose and enforce a US federal regulation on a non-US airport. Thus, it became abundantly clear that charge removal would have to occur—not from outside the airplane tank—but from inside the airplane composite fuel tank as part of airplane fuel tank design.

A reasonable question could be asked at this point: Why was the above-stated situation never a problem on airplanes built prior to the 787 Dreamliner?

The answer to this lies in the fact that fuel tanks installed on the 777, 737MAX, 737NG, 747-8, and other similar BCA series of passenger airplanes were aluminum (Al 2024). The Al metallic fuel tanks are inherently physically electrically connected to the all-metal airplane primary support structure by way of strong metal fasteners. The primary airplane support structure serves a dual role.

While the basic aluminum structure is the backbone of the Boeing commercial jet airplane, it also serves very well as the airplane's zero-volt reference system ground.

When the high charge-laden Jet A airplane fuel enters the Al 2024 metal fuel tank, the attachment of those tanks to airplane primary structure provides an excellent ready-made charge bleed-off path to airplane system ground for the high positive electric charge in the fuel tank (see Figures 16.2(a) and 16.2(b)).

It is the force of positive electric charge attraction to airplane system ground that is the force behind the removal of charge from the tank, resulting in the prevention of sparking taking place inside the fuel tank [1]. Thankfully, this charge removal occurs rapidly. Thus, fuel system engineers, prior to the Carbon Fiber Reinforced Plastic (CFRP) composite in the 787 Dreamliner and fuel tank, never had to solve the problem of high charge in the fuel tank.

However, not so with the 787 Dreamliner.

The introduction of the lighter-weight, nonaluminum CFRP in the BCA 787 airplane fuel tanks unfortunately required the fuel tank inner lining to be covered in a noncorrosive chemistry of paint. Why? Because Jet A fuel is corrosive to CFRP. To mitigate this problem, a dielectric multilayer paint stack-up was applied to the inside wall of the fuel tank.

Problem solved, right? Wrong.

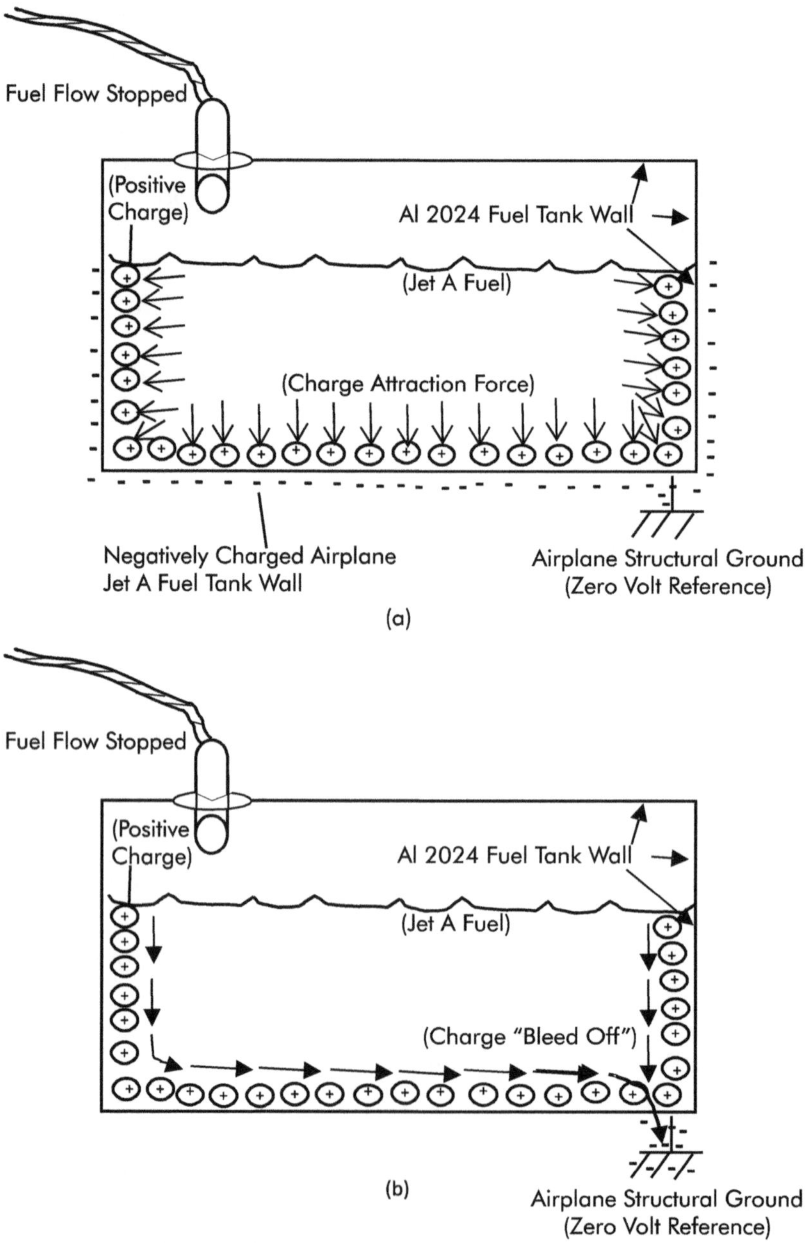

Figure 16.2 Illustration of (a) electrostatic charge drawn to Al 2024 airplane tank wall, (b) electrostatic charge bleeding out of Al 2024 airplane tank to airplane electrical ground, (c) electrostatic charge collecting along inner wall of CFRP airplane tank, and (d) stationary electrostatic charge along inner wall of CFRP airplane tank (no charge bleed off).

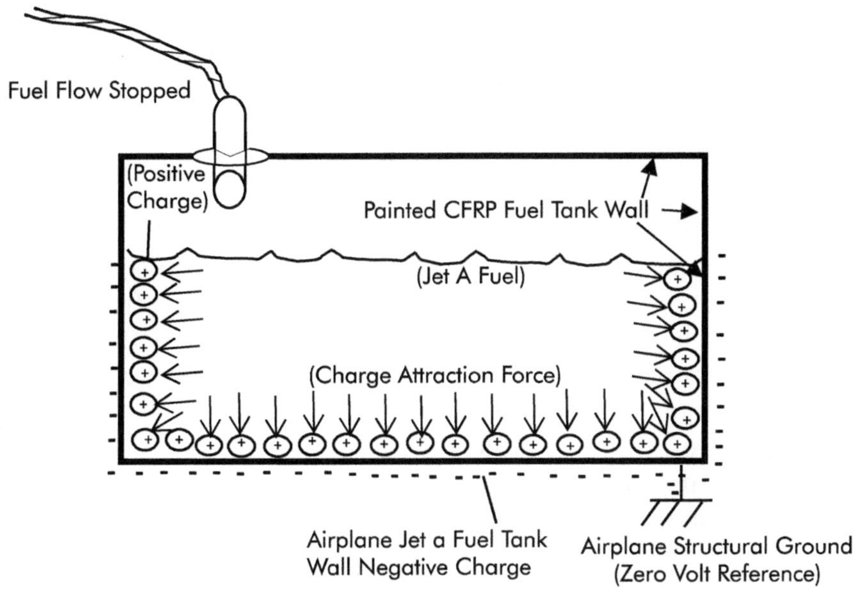

(c)

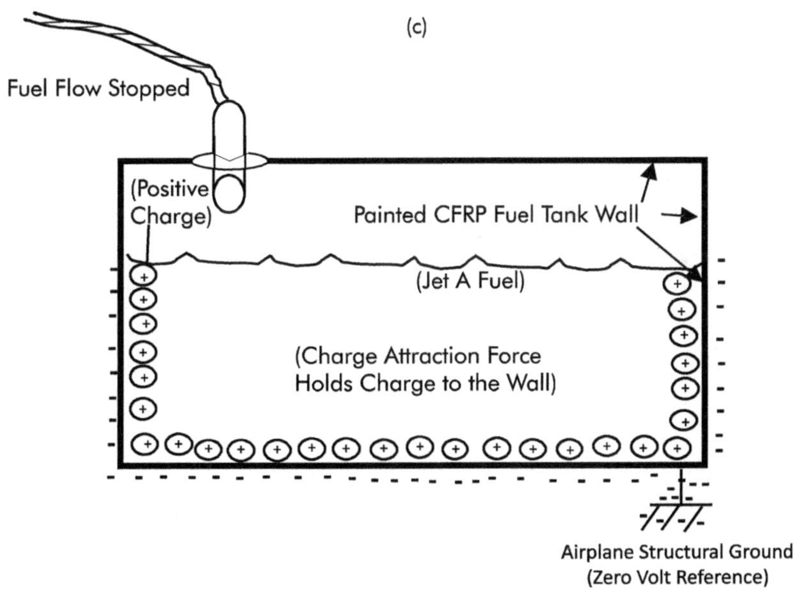

(d)

Figure 16.2 *(Cont.)*

Unfortunately, while the paint protected the inner CFRP fuel tank walls from corrosion, it also had the unwanted side effect of acting as an insulative barrier to the charge bleed-off function. The charge, while drawn to the fuel tank inner wall, would simply be held down to the tank wall by same electrical attraction force (see Figure 16.2(c, d)).

So, now what? The devised solution for the manufacture of Boeing airplane fuel tanks was, and still is, Boeing proprietary. As such, this solution cannot be discussed without permission from Boeing. But suffice it to say that the solution, while quite effective, had the unfortunate consequence of making airplane tanks more labor-intensive and, thus, more expensive to produce.

The bottom line was that the next challenge for the PD group (specifically, me) was to find a different solution to the charge accumulation problem in the CFRP airplane fuel tank. The aim would still be to remove the high amount of electric charge after refilling. But the solution found had to be one that was much more cost-effective (i.e., less expensive) to implement in production.

As a result, we tested a number of materials for the inner wall lining in the electrostatic physics lab. Unfortunately, each proposed material proved to be too electrically insulative, that is, too resistant to allow charge in the fuel to pass through it (like resistor) to airplane electrical ground. It soon became apparent that the only solution would be to find (or develop) a suitable noncorrosive inner fuel tank wall lining material. I regret to say that, as of my June 5, 2020 retirement from the Boeing Company, a suitable airplane fuel tank lining material was not found by me (the group's senior principal investigator) or any other member of the team. However, the upside of this story is that, through encountering this airplane fuel tank challenge, I gained tremendous exposure and experience to the world of airplane fuel-based electrostatics. The experience enhanced my understanding of fuel tank electrostatic problems immensely.

16.1.2 Human-Touch ESD on Electronic Printed Circuit Boards

16.1.2.1 Problem

In the 1970s, fully IC-populated but failing printed circuit boards (PCBs) were being delivered to certain expectant customers of those boards. The problem was that the PCBs had failed TTL technology integrated circuits (ICs) installed on them. Failing IC-populated PCBs were coming off the manufacturing assembly line at far too high a number (40%) for the company to be profitable. Furthermore, the company had no idea why this was happening. As such, an investigation was begun to find out the cause and find a suitable problem remedy.

During the investigation, it was discovered that assembly line personnel were unknowingly generating high levels of electrostatic charge on their person across the workday. In the process of handling the PCBs, the assembly line workers were, at the same time, unknowingly introducing electrostatic charge at the workstations by touching the boards. The inadvertent dissipation effectively "killed" the ICs on the cards both before and after acceptance testing. Neither the human eye nor the human touch could discern what was happening as it was happening.

16.1.2.2 Explanation

The situation was much like when a person walks across a living room carpet in dry winter air, unknowingly generating a considerably high charge on their person. Then, the person touches a metal doorknob. The result is a considerably uncomfortable "snap" upon coming in contact with the metal doorknob. That common wintertime experience is in essence the same thing that was happening in the PCB assembly factories. The difference was that for the case of the PCB assembly factories the discharge that was taking place was at a level so low that it could not be felt or seen. Yet, ICs were still leaving the factory having been damaged as confirmed at PCB acceptance testing. It was conjectured that electrostatic discharge (ESD) had to be the culprit.

In time, the investigators found that factory workers were unknowingly generating high levels of electric charge on their person both before and during their shift.

While handling the printed circuit cards, they were unknowingly dissipating their self-generated electric charge across the day on the IC-populated PCBs. Workers would subsequently take the newly assembled PCB cards over to the acceptance test station for testing, and a large number of PCB assembly cards would fail acceptance testing.

So, with the problem culprit found, the next step was to find a remedy to the problem.

16.1.2.3 Solution

Several factors contributed to the ESD problem, and a number of problem mitigations were recommended. A two-step approach was taken. The initial step was to first prevent charge-generation from occurring as much as possible. The second step was also to remove any stray electrostatic charge before the assembly worker could touch a PCB.

One step taken was to raise factory PCB assembly area humidity levels (e.g., to 60%). However, while this step helped, it did not fully mitigate the problem.

The second measure was to have all factory personnel avoid wearing charge-generating clothing materials to work altogether. Thus, what had to be identified was charge-generating clothing materials. Wool was found to be a key charge generation clothing material (sweaters, socks, gloves, pants, hats, etc.). As a result, wool clothing was immediately banned from being worn in the factory work area, particularly, the factory PCB assembly floor.

In its place, cotton-based clothing proved to be the least static-charge generating clothing material, and factory workers were permitted to wear cotton-based clothing to work. By doing this, the static charge generation was even further reduced. Other prevention measures were taken as well.

The donning of full-length (i.e., neck to calf) conductive smocks with copper woven into the material were mandatory for all factory PCB assembly personnel to wear while working on the PCB assembly floor. The conductive metal woven into the smock fabric helped to dissipate static charge that was inadvertently and unknowingly being generated on the human body across the day.

Investigators also found that accumulated human body static charge could be most effectively "bled off" the working person before the PCB assembly worker would begin to solder IC chips to the PCB and handle the boards thereafter. Bleeding off human-generated charge would be effectively accomplished by means of the worker wearing an ESD wrist strap. See Figure 16.3 [1]. The ESD wrist strap works essentially as described in the following manner.

The accumulated charge on the human body is bled away by way of the electrically connected wrist strap. The wrist strap end must be touching the human skin and connected in series to a long, coiled conductive wire. The conductive wire is then also electrically tied (in series) to a 1M-Ω resistor with the resistor next tied to a shorter conductive wire that is then finally again in-series attached to a workstation earth-ground point. See Figure 16.3 [1]. The workstation configuration successfully:

1. Bleeds the generated charge from the person;
2. Sends the charge down the wire and away from the PCB boards being hand-soldered or handled;
3. Finally dissipates the charge through the 1M-Ω resistor.

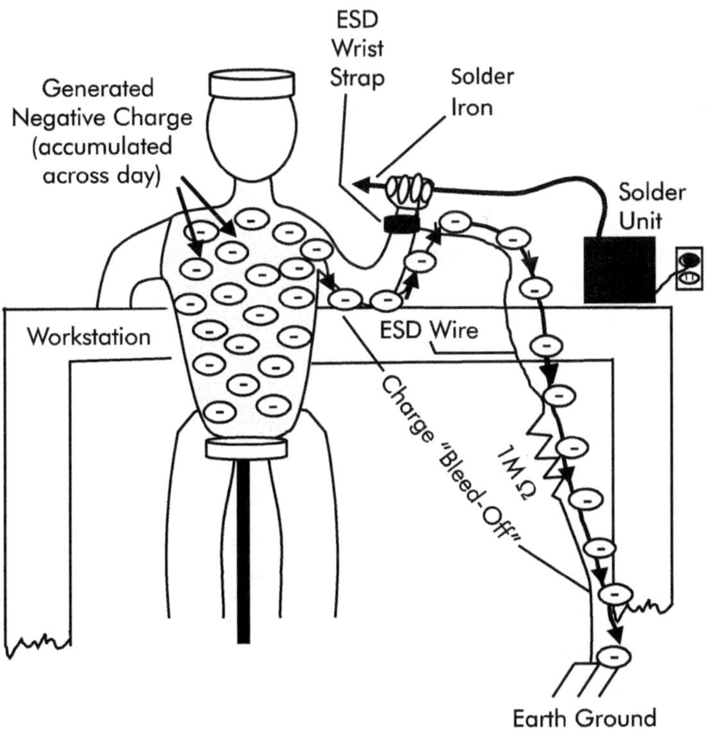

Figure 16.3 Illustration of electrostatic charge bleed-off by use of an ESD wrist strap and conductive wire coil.

When the electrostatic charge removal (i.e., bleed off) system was also implemented at all PCB factory workstations (1970s), assembled and IC-populated PCB's began passing final acceptance testing. The immediate results found were that PCB passed acceptance test at 95% of the boards or better. The above-described measures all taken together in the factory PCB assembly floors virtually eliminated the ESD problem for PCB manufacturers.

While I (the author) worked at the Northrop-Grumman Corporation (NGC) in Huntsville, AL from 2000 to 2003, assembly card stations and employees all gladly implemented the previously described measures. Everyone entering the factory area, top-level managers on down, wore the conductive smocks and could only look at the work being performed. As a result, ESD-damage to printed circuit cards and RF chip modules were no longer a factor.

The next static charge case involves no human interaction at all.

16.1.3 P-Static Noise Captured in In-flight Commercial Jet Airplane HF Radios

This case involves high levels of electrostatic generation called precipitation-static (or p-static). It involves only airplanes and no human interaction.

16.1.3.1 Problem

The symptom of the problem was high frequency radio (HF) static being heard in the airplane HF radio receiver by the pilot and first officer while the Boeing prototype airplane was in-flight. The 1950s Boeing jet prototype (i.e., the Boeing 386-80 jet airplane) HF radio could communicate quite well with other HF radios while the airplane was on the ground. But the HF radio static noise simply would not let the pilot and his first officer (FO, a.k.a., copilot) communicate either with ground-based HF radios or other in-flight HF airplane radios. Of course, it took time for the Stanford Research Institute (SRI) investigator(s) to figure out the source of the problem.

The in-band noise was so high that you could not tell whether or not the transmitting HF radio was functioning properly. Pilots must be able to communicate in-flight to the ground tower in order to have successful take-off, ascent, cruise, descent, and landing phases of airplane flight. So, how was the source of the HF radio noise problem identified and finally mitigated?

Problem-Source Identified SRI investigator Joseph E. Nanovicz and SRI colleagues performed interesting ground-based experiments.

16.1.3.2 Discovering the Cause(s)

The conducted experiment is briefly described. A gun bullet was fired down the length of a long dielectric (i.e., nonmetallic) cylinder. The cylinder contained floating ice crystals and subzero cooled air.

Outside the dielectric cylinder, an old HF-band antenna was set up with the antenna feedline connected to the RF input port of an old 1950s-style analog spectrum analyzer/receiver with the span of the HF band instrument set from 2.0 to 22.0 MHz. In essence, as the bullet traveled down the cylinder, it collided with in-path ice crystals. The result was that an HF-band noise was being momentarily radiated/emitted from the cylinder as detected by the spectrum analyzer.

The experimenters postulated that, as the bullets were colliding with the floating ice crystals, the bullets had to be collecting static charge on them by way of tribo-electric effect while traveling and, hence, emitting noise in the 2.0 to 22.0 MHz HF band. Thus, it was concluded that what was happening in the cylinder had to also be occurring on the 386-80 prototype jet airplane in-flight, too.

To confirm, the SRI researchers knew that ice crystals form at subzero altitudes above dark, rainy thunderstorm clouds. So, the arrangement was made to fly the 386-80 prototype jet airplane at subzero altitudes above thunderstorm clouds. The result was extremely strong 2.0 to 20.0 MHz HF-band noise could be heard in the HF radio headsets by both the pilot and his copilot in flight. The noise amplitudes heard were well above any amplitudes from transmitted HF voice signals sent from the ground to the test airplane (as confirmed by preflight ground testing).

The full truth was that the problem was multilayered. The tribo-electric effect was a key part of the cause but not the full story.

What was not understood at the time of the effort was that the paint on the orange-colored prototype test airplane was a nonconductive, dielectric material. That dielectric material collected charge on the outside of the fuselage, flowed it down the fuselage to the trailing edges of the airplane wings, and large random sparks could be seen. The spark emitted the HF-band noise. The airplane HF-band antenna was capturing that noise and sending it down the transmission line to the HF-band receiver, effectively swamping out HF-band voice signals also being transmitted to the HF antenna from the ground as could be heard by pilot/first officer headsets. HF voice communication simply could not be heard in the flight deck.

Researchers also speculated that any airborne particles, such as snowflakes, dust, and volcano smoke, could also bombard the airplane at enormously high rates during airplane take-off, ascent, cruise, descent, or landing, too. Each time these particulates strike the outside metal airplane skin, electric charge is generated on the painted airplane exterior and HF-band noise was generated and radiated.

If you can imagine multiplying the cumulative triboelectric effect produced from hundreds of thousands (or, perhaps, even millions) of these tiny particulates bombarding the airplane in-flight, it is easy to then understand how enormously high amounts of electric charge can get collected and stored on the airplane fuselage.

16.1.3.3 Detailed Electrostatic Physics Involved

It is important to understand that the deposited charge will not simply ride on the airplane. Instead, a force of electrostatic attraction (i.e., Coulomb's Law) takes place between the charged particles on the airplane and the air molecules from which they were taken [1]. The airplane flies past those oppositely-charged air molecules, and, thus, the negatively-charged particles will travel across the 41-section airplane skin of the fuselage to both of the wings. They are then pulled off the wing trailing edge to recombine with the oppositely-charge air molecules at the wing's trailing edge and make the net charge zero again.

There is, however, a last part of the process.

When the charge particles jump off the wing's trailing edge, an electric arc forms. This arc is the problem.

The arc is the result of the dielectric breakdown voltage of the air between the air and the oppositely-charge air molecules being exceeded. This "snap" occurs repeatedly. The snap radiates and propagates a very wideband electromagnetic wideband noise that also includes as a subset the 2.0 to 22.0 MHz HF radio band.

16.1.3.4 Other Commercial Airplane Receivers and Radios Affects by P-Static

The frequency band of the radiated noise also spans the same receiver channel frequency range as the navigational automated direction finder (ADF, receive-only frequency range: 0.190 MHz to 1.995 MHz), the navigational VHF omnidirectional receiver (VOR range: 108.0 to 118.0 MHz), and the very high frequency radio (VHF range: 118.0 to 137.0 MHz). The radiated noise in each of these cases is subsequently captured by the associated radio/receiver antennas and sent down the associated transmission lines on the airplanes.

The noise going into those receivers is subsequently demodulated into audio band frequencies. The noise then goes down the audio frequency transmission lines and to the pilot/first-officer headsets. The baseband being picked off the carrier frequency is the radio static noise. This noisy baseband is the noise that is heard in the pilot/first officer headsets.

Of the four radios/receivers mentioned, the magnitude of the ADF band noise is the strongest of the four commercial airplane bands. It can be high enough to not permit the ADF to detect the much-needed presence of the nondirectional beacon (NDB) signal in the potentially detectable vicinity of the airplane. It, too, must be dealt with.

The next strongest airplane radio band is the HF radio frequency band. Across the 2.0 to 30.0 MHz, the captured p-static noise, if not dealt with, tends to overwhelm the HF channels, and HF band communication cannot be established.

The p-static noise amplitudes in the VOR and VHF bands, most fortunately, tend to roll off appreciably and at such low amplitude(s) that no RF interference is detected by the VOR navigational receiver (108.0 to 118.0 MHz) or the copilot operator of the VHF radio (118.0 to 136.0 MHz).

So, then, if the result is that captured in-band noise is high enough such that ADF receivers cannot detect the navigational NDB in the intended proximity, how then is the pilot expected to receive navigation from the ADF to steer the airplane to its intended direction? One answer is, of course, to use the VOR navigation receiver instead. Likewise, if the result is that the amplitude of the captured in-band HF noise is high enough to swamp out incoming HF radio voice signals, how are pilots expected to receive instructions for safe airplane approach and landing? A voice communication alternative is to use either the on-board VHF band or UHF band radios. Regardless, the FAA requirement is that airplane must be able to use the ADF receiver for navigation and the HF radio for voice communications.

Solution 1

It was discovered that specifically designed static dischargers (a.k.a., p-static "wicks") could be designed and positioned on the trailing edge of the airplane wing to help minimize the magnitude (and the impact) of p-static noise. Several discharger designs were made and tested both in laboratory settings and on the airplane. Those "sticks" that you see on the trailing edge of either the left or right airplane wing, as you the passenger are looking out the window at the wing's trailing edge on your side of the plane, are those p-static discharger "wicks."

Solution 2

Unfortunately, on the pre-707 (Boeing 386-80) jet prototype, it was found that designing optimal p-static wicks did not eliminate all the airplane's radiated p-static HF radio and ADF receiver navigational noise. As described earlier, early dielectric airplane paint tended to exacerbate the p-static noise. As a result, early Boeing airplane fuselages had very little paint on them. As such, the ADF band and HF band noise was minimized such that both systems worked well on the airplane in-flight. Unfortunately, another layer of the problem was found.

Solution 3

It was also found that airplane p-static noise could be reduced further by electrically joining (bonding) all the major aluminum metal primary structure pieces. This could be done through the use of "bond jumpers" electrically attached between each the major aluminum airplane structure piece on the fuselage and airplane wingspan. Without the bond jumpers, physical and, more importantly, electrical gaps—that is, metal parts separated by an air dielectric—were found to exist between the major Al primary structure pieces. The result was that electric charge collected more on some major airplane structure pieces than on others.

Thus, when the potential difference between the structural pieces exceeded the dielectric breakdown voltage of the air gap, arcing would begin. With the arcing, came ADF and HF band noise radiation. With radiation came ADF and HF band interference that would take place in the respective systems and create problems for the pilots. Bond jumpers placed between major airplane structural points served to bring down p-static radiation on the airplane quite successfully.

With all three p-static noise mitigation solutions implemented, the p-static noise interference problem in ADF receivers and HF radios all but went away for many years. However, in time airline desire to fully paint airplane fuselages and wings returned in the effort to make a company's own airplane fleet stand out to customers. Fortunately, for a time, it had been remembered that dielectric paint would cause the p-static noise to return. In its stead, metallic based paint was used to cover fuselages and that solution worked quite well.

Unfortunately, in time airlines wanted to have the cost of a new airplane reduced. So, back came the lesser expensive dielectric paint and, to a degree, so did p-static. However, the level found in ADF receivers and HF-band radios was not enough to prohibit successful functioning of the airplane's ADF receiver and HF-band radio system. So, the new dielectric paint stayed on Boeing Commercial Airplanes for several decades.

16.1.3.5 P-Static Noise Returning to the BCA Airplane

In the year 2000 and beyond, airplanes also began to be seen by certain companies such as Walt Disney and certain soft drink companies as advertising and promotional opportunities. As such, those companies paid airlines good money to place very large acreage decals/appliques on fuselages completely covering the paint. After having decades of no HF radio noise, it was found that HF radio noise returned to pilot headsets. Boeing was then asked to investigate why this was so.

In the investigation, it turned out that Boeing indeed delivered painted airplanes that passed p-static flight testing. However, the airlines applied paper decals and appliques over the paint and on the entire fuselage, and the HF radio noise returned. In fact, the resulting noise saturated on-board HF receiver radios in-flight. In 20/20 hindsight, one might think that the cause for the problem should have been obvious to Boeing investigators. However, it is important to know that the older engineers, who had mitigated the earlier problem, had retired. A whole new generation of airplane engineers had to figure out the problem all over again.

In short, it was found that, when the decorative appliques/decals were removed, the HF radio noise went away. (Note: This was when I became involved.) Then, the question became why it was that these decal/applique materials were introducing wideband noise into on-board HF commercial radios. It was postulated and confirmed that the decal/applique materials were dielectric in nature.

As such, it was concluded by the senior principal investigator that was performing the investigation (i.e., myself) that the decal/applique materials were not effectively removing the accumulating electrostatic charge (i.e., p-static charge) from the in-flight airplane exterior. Because large advertising dollars were being received by the airlines for logo display on airplane fuselages, there was a strong push from a variety of companies to have Boeing solve this problem. As such, large area applique/decals were not going away.

The next technical challenge was for applique/decal manufacturers to develop decal/applique materials that would effectively dissipate the high accumulated p-static charge while in-flight, and, thus, eliminate the HF-band radio noise in pilot and first-officer headsets. The relative ease and swiftness of applying and later removing appliques/decals made decal/applique logos highly desirable to airlines.

In time, new suitable decal/applique materials were found/developed that could effectively dissipate p-static charge from airplane jets in-flight. However, because no documentation citing how effectively the applique/decal could dissipate electrostatic charge was made available for review to the senior principal investigator, p-static laboratory tests were invented, developed, and used. They were devised to observe and measure the ability of these materials to dissipate accumulating p-static charge. In the end, two decal/applique materials at Boeing were identified as being all to absorb the p-static charge directly to the airplane conductive airplane skin.

Hence, the p-static problem was solved, and the investigation was over.

16.1.4 Preventing Electrostatic Shock to Airline Ground Crew Post-Landing of Commercial/Military Jet Aircraft

As discussed in the previous section, an in-flight jet airplane accumulates considerable electrostatic charge. The accumulated p-static charge does not simply dissipate off the fuselage after landing and being parked. It must be removed and safely, too. The charge is high enough in magnitude to potentially kill or injure a ground crew member that prematurely touches the airplane body post-flight. Thus, a method for safely and swiftly removing this extremely high residual static charge from the fuselage had to be devised.

The answer: Borrowing again from Coulomb's law, "bleeding," that is, bleeding the residual charge to tarmac earth ground. An illustration for how this is done is shown in Figure 16.4.

Now the next question to ask is why there remains an incredibly large residual static charge on the body of an airplane after flight.

16.1.4.1 Cause for Residual Charge

Believe it or not, an airplane essentially takes on the characteristics of a large electrical capacitor as charge accumulates on it in-flight. Recall that a capacitor is in essence two conductors separated by an insulator. Conductor 1 is the airplane's aluminum metal outer skin. On top of the conductive skin is an insulative combination of

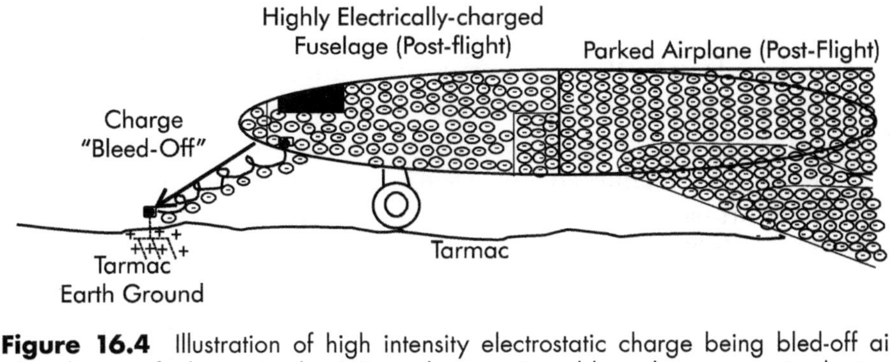

Figure 16.4 Illustration of high intensity electrostatic charge being bled-off airplane by use of a large-sized AWG conductive wire cable and an appropriately-sized electrical resistor.

dielectric materials. The combination will either be a standard combination (or stack) of airplane paint layers or the combination of a few layers of airplane paint with the dielectric decal/applique placed on top of it. A final transparent or clearcoat (i.e., no pigment) paint is the final layer.

The conductor 2 will no doubt surprise you.

Conductor 2 is the cumulative precipitation static charge that the airplane accumulates in-flight. The charge sits on the airplane skin because it is essentially electrons being held to the top of the dielectric paint/decal/clearcoat by way of the electrostatic force of attraction (again Coulomb's law) taking place between top-sitting negative electrons and the positive charge of the conductive airplane skin.

Now that we know how the airplane capacitor is comprised, the solution to the dangerously post-flight high static charge problem will be explained.

16.1.4.2 Solution

The solution for this problem is amazingly simple. Just as you have to bleed off charge of a human body before the human worker can hold of an IC-populated printed circuit board, so also must this high charge be bled off from off the aircraft, be it commercial or military, before passengers and flight crew can deplane and ground crews can perform post-flight maintenance.

On every commercial and military airplane, a charge bleed-off connection termination point exists on the front of the airplane. A large-sized (AWG) conductive cable is electrically connected (i.e., clamped) to it at one end of the cable. The opposite cable end is likewise electrically (and securely) connected to a tarmac-based earth-ground termination point located nearby. With both cable ends securely and electrically connected, the charge on the fuselage bleeds off (discharged from) the aircraft quickly and completely. The bleed-off cable is kept in place for a long time, that time being approximately 5τ. (Note: τ is the R-C time constant, C is the airplane capacitance, R is the inherent resistance of the bleed-off cable.)

After the time of 5τ is complete, the cable is disconnected at both ends and removed. Following cable removal, airline ground crews are free to work and complete post-flight maintenance. Passengers can be let out of the aircraft safely.

16.1.5 Discharging High-Magnitude ESD from US Army Cargo Helicopters Before Connecting Tactical Cargo to Hovering Transport Helicopters on the Battlefield

Helicopter platforms rapidly generate huge amounts of electrostatic charge after leaving a ground departure point. The cause for the charge accumulation is the same triboelectric effect, occurring from the helicopter blades that are interacting with the surrounding air (dust particulates colliding with the helicopter rotor blades also contribute considerably to the electrostatic charge generation).

After the helicopter arrives at its designated cargo delivery destination, it hovers over the point with the cargo hanging down from the helicopter on a line. As ground personnel attempt to detach the cargo from the line, they experience large jolts of electrostatic charge, knocking them to the ground to the point of medical attention being required afterward.

How did the US Army mitigate this problem?

16.1.5.1 Solution

The US Army Corps of Engineers ascertained that the high accumulated charge had to be bled off the combination of the helicopter, transport chain, and the attached load before ground personnel could detach and remove the cargo. Before cargo removal, US Army ground troops drive a large 2' to 3' metal (iron) stake into the ground below the hovering craft. The pilot next drops a conductive metal chain from the helicopter to the Earth. The chain is to touch the iron stake so that the electrostatic charge can be removed and go straight to earth ground. In training films, you can see a visible spark at the point where the hanging metal chain touches the stake. This is the moment that static charge is being removed (or bled away) from the helicopter. Afterward, the cargo is lowered to the ground and safely removed by the ground crew. See Figure 16.5.

16.2 ANECDOTES IN MAGNETOSTATICS

In April 2009, I encountered a malfunctioning RF test set with a truly baffling problem.

The matter was first brought to the attention of the other RF group teammates, and they could not solve it. Finally, I was asked to

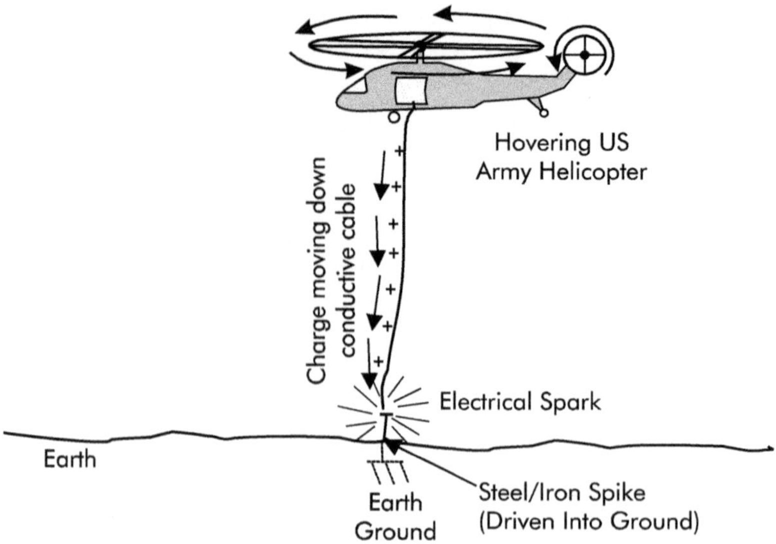

Figure 16.5 Removal of helicopter-generated static charge.

look at it. To the surprise of my RF engineering colleagues, I saw the problem and quickly solved it.

16.2.1 VHF Radio Test Set Fails to Simultaneously Transmit VHF Voice or Data and Receive VHF-Band Satellite Messages As Designed

The objective of this 2009 B52H engineering project was to provide the bomber with the added capability of both receiving VHF-band satellite intelligence data while simultaneously conducting two-way in-flight VHF voice communications at any time. The RF circuitry for this added capability had been designed by an otherwise very capable junior electrical engineer (EE) in the System Engineering organization and not someone in our Boeing-Wichita RF group. Unfortunately, what the junior EE had inadvertently not considered in his RF design was the potential for the Biot-Savart Law to arise in a most negative manner.

16.2.1.1 Path to Solution: New RF Circuit Test Symptoms

What made this situation truly disconcerting was the fact that the program directed the new RF circuit design to be installed on the bomber immediately after initial design (i.e., before being thoroughly tested prior to aircraft installation). Fortunately, the system engineering team had the foresight to install an exact duplicate of the new bomber RF circuit in an 19" test rack. The SIL test rack was located in the Boeing-Wichita building 190Z SIL. It was during some SIL test set runs (using the new RF circuit) that a troubling operational problem was uncovered.

The System Engineering team fortunately also had the foresight to exercise every new RF circuit function that was to be performed on the B52H bomber airplane. It was during the SIL system test set "ring-out" that the unforeseen problem reared its ugly head—the test set would not receive downlinked intelligence data from an orbiting satellite while simultaneously performing two-way VHF radio communications.

Simply put, the test set operator was trying to transmit VHF voice or digital data while simultaneously trying to downlink the intelligence satellite data. Sometimes, the effort was successful. However, other times, the result was that the received data was complete, unintelligible garbage. The cause for this had to be discovered before scheduled capability delivery and time was running out.

To their credit, the senior lab test engineer and the junior electrical engineer (i.e., the young man that designed the new RF circuitry) did their best to isolate the problem. But, after many days, they simply could not do it. It was after they gave up that they contacted certain of our Boeing-Wichita RF group for help. Unfortunately, none of my other RF engineering colleagues could solve the problem. Finally, I volunteered to go help. I then found the problem. What follows is a description of the steps I undertook to isolate the problem.

16.2.1.2 First Steps

I knew that the first step was to replicate the problem and examine the RF test circuitry response. The senior SIL lab test engineer and I did just that. After a few successful (and uneventful) runs, the problem reared its head again. We were successful in duplicating the troubling RF test set symptoms. After documenting them, I took them back to my desk for further study and contemplation.

It was interesting to note that the symptoms did not appear frequently. However, they did so often enough for the test set engineer and I to clearly observe them. I next asked the junior EE design engineer for a paper copy of the VHF circuitry (see Figure 16.6).

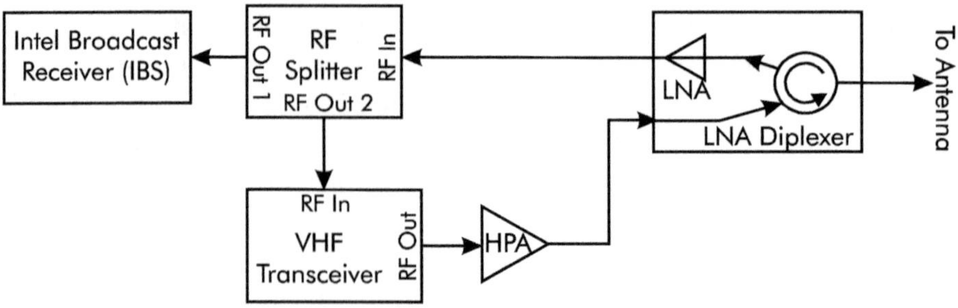

Figure 16.6 RF test set circuitry added to the B52H System Integration Lab (SIL) test set.

I became quite proficient in designing, analyzing, and evaluating RF circuitry. This came from years of professional RF engineering design, testing, and installation experience. As a result, I was able to interpret and completely understand the RF schematic circuitry of the figure (Figure 16.6) quickly. The source of the VHF-band interference problem all but leapt out at me from the diagram.

However, before revealing my suspicion to anyone, I thought it well to thoroughly study the corresponding B52's VHF ARC-210 Warrior radio transceiver specification. I studied the particular specification paragraphs that pertained to the expected radio spurious emissions. In them, I found the maximum expected output of VHF-band spurious emissions that could possibly be emitted from the radio's RF output port. I then studied the new RF circuitry. In a few minutes, I figured out what I thought was the source of the problem.

The transceiver out-of-band spurious emissions coming out of the VHF radio during voice transmission were coupling into the VHF intelligence receiver RF port by way of the designer-selected LNA diplexer component. See Figure 16.7(a).

When the Intelligence Broadcast Service (IBS) satellite randomly downlinks VHF-band intel data at the same time two-way VHF voice communications were occurring, the much smaller IBS data would be completely overwhelmed by the far larger magnitude H-field coupled voice communication transmit signal. Sometimes both would arrive at the Intelligence Broadcast Receiver (IBR) RF input port at the same time. See Figure 16.7(b).

The source of the problem was finally identified and isolated (see Figure 16.7(b) diagram markings).

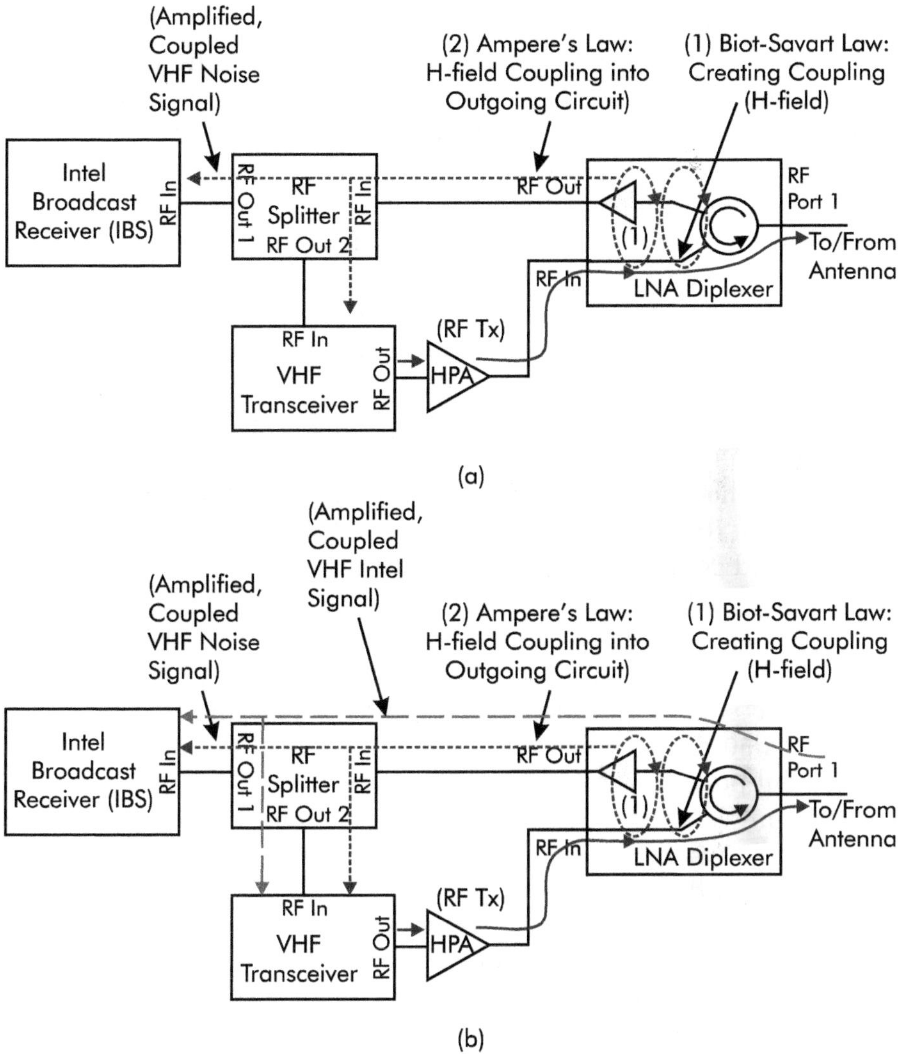

Figure 16.7 (a) Biot-Savart law at work: H-field created from VHF voice transmission signal. (b) Ampere's law at work: coupled H-field (noise current) of VHF transmission going to VHF intelligence broadcast receiver RF input port.

The LNA/diplexer simply had to be the source of the VHF interference problem. Indeed, it was. Here is why.

See IBS RF signal path in Figure 16.7(b). When VHF voice communication (transmission) starts, the VHF transmit voice signal goes from the VHF transceiver 100W RF output port to the RF input port of the 30 dB VHF-band high power amplifier (HPA), getting amplified 1,000 times more in power. Next, the very high-powered VHF signal then enters the LNA/diplexer through the RF In port. From within the LNA/diplexer, the amplified voice communication signal exits the LNA RF component through RF port 1, finally radiating out the VHF antenna. This is the intended design.

However, not all the amplified voice communication signals go out LNA/diplexer RF port 1. A portion of it gets converted to an H-field (Biot-Savart law) inside the diplexer. See again Figure 16.7(a) (for a review of Biot-Savart Law, see Section 9.2) [1]. The H-field inside the diplexer gets coupled (via Ampere's law; see Section 9.3) into the low noise amplifier (LNA) RF input port conductor [1]. The coupled signal current is amplified again by another 25 dB while passing through the LNA. The high-power signal passes out the LNA/diplexer and through the RF splitter to arrive at the RF IN port of the VHF IBR. This H-field coupled signal will enter the VHF intelligence receiver every time the RF circuit operator attempts to transmit a VHF signal out of the transceiver.

The pit that the junior design engineer fell into was, believe it or not, the supplier's LNA/diplexer documentation. It claimed that there was 50-dB RF isolation (i.e., loss) between the RF in port and the RF out port. However, to save the manufacturer from lawsuit, an annotation was written in very tiny font. It said that the 50 dB of RF isolation between the RF port occurs when dc power going to the LNA/diplexer device is off.

Now, I ask you: Who in the world writes an utterly useless device specification statement like that?

When implemented in an actual circuit, the only time the RF isolation would be needed was when the LNA/diplexer was powered on. Yet, the device manufacturer offered no dB value for RF isolation between those RF ports for when the device was powered on. To this day, I have never understood why the device manufacturer allowed this specification to be printed. In my own 37 years of RF engineering, no active RF device is used in an RF design circuit with the dc power to the device off.

However, the unavoidable fact was that IBS satellite signal could be downlinked from the satellite at any time without warning. There was no way the test set operator could control or predict the coming situation. So, when VHF signal 2 was boosted up another 25 dB by the LNA in the diplexer, in the end, the VHF voice communication transmission was boosted so much higher than the IBS signal that the satellite downlink signal was completely overwhelmed by the transmitting voice communication signal. As such, VHF satellite intelligence receiver could neither detect nor effectively demodulate the incoming amplified Signal #2 carrier. Hence, whatever demodulated baseband was received in the VHF intelligence receiver ended up with unintelligible noise (i.e., complete digital garbage).

This was how the RF design circuit culprit was identified and isolated.

16.2.1.3 Last Steps: Conclusions Drawn

The selected LNA/diplexer did not have sufficient RF isolation between RF ports 2 and 3 due to a misleading component supplier document. I myself went to the RF lab and measured the RF isolation between the two ports in the LNA/diplexer (and, of course, with the device's dc power on). The RF isolation was not 50 dB—it was 35 dB. This level of the needed RF attenuation was not going to reduce the coupled RF signal low sufficiently to mitigate the RF interference in the system.

In the effort to save the basic design, I subsequently did a thorough internet search to find a replacement LNA/diplexer. Unfortunately, no company as of April 2009 could supply the same device with the required RF isolation performance between the two ports. Thus, unfortunately, too, the flaw was left in the design as it was already installed in the bomber airplane.

The important lesson to take away from this unfortunate calamity is to be sure to not only read all specification notes pertaining to an RF device carefully before selecting it but also to consult with an experienced RF engineer, just to be on the safe side.

16.3 ANECDOTES IN ELECTROMAGNETICS

16.3.1 Case of the Confirmation of 10X-Extended Range of the Manned Control Station-to-Missile Launch VHF Signal

Following the 1990 Gulf War, the US Army recognized that the tactical deployment of a certain missile weapon system (it will remain unnamed; it is still deployed around the world) had to be changed in order for launched surface-to-air missiles (SAMs) to more effectively shoot down incoming enemy tactical ballistic missile (TBM) threats. It is important, however, to first understand that this weapon system was originally designed to shoot down air-breathing threats (ABTs) and not TBMs. Regardless, it is typical for the US Army to stretch a weapon systems launch range capability beyond the original limit of the system.

However, it was hoped by US Army tacticians that, if this system's missile launch stations could be deployed closer to the target threat, a higher threat missile kill rate could be achieved. What would need to be changed is the separation distance between the missile launcher and the manned launch control station. The new needed distance was 10× the distance of the original weapon system range.

What must also be understood is that no one at the system project office was certain that successful missile launch communications to the extended distance could be achieved. Nonetheless, an outdoor VHF radiation communication test was developed, planned, and executed at the White Sands Missile Range (WSMR, NM). An illustration of the test scenario is depicted in Figure 16.8.

What made this February 1992 assignment quite interesting was that the outdoor range test had already been conducted. The test results were collected, and the project office personnel said they could not interpret the success or failure of the test from the results. Not even my otherwise very clever manager could perform this test results evaluation. So, he handed the problem over to me.

After Wayne, my technical manager at the time, explained the events that led up to the current situation, he told me that I would not get to first look at the test result numbers. He asked that I make a completely unbiased determination of the test outcome. When I was finished with my evaluation/analysis, I would then get to see the test results.

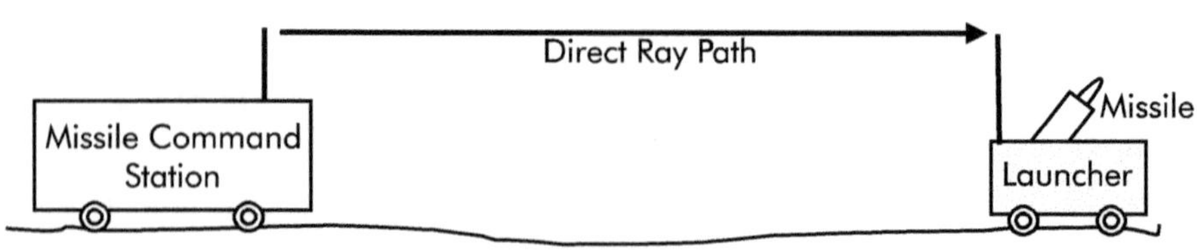

Figure 16.8 Basic illustration of the outdoor test set-up.

16.3.1.1 Approach Taken

Unfortunately, those at the project office that devised the test had not made a predetermined success criteria (i.e., pass/failure) in advance of the actual test exercise. So, to approach the problem logically, I pretended that the outdoor test had yet to be conducted, and I did the most logical thing that I would do if I were the test director/conductor. I would determine the test success criteria numerically.

16.3.1.2 Steps Taken

First, I performed a basic Friis equation-based numerical prediction of what the test results should be. Next, I had to research what the RF losses should be experienced in the test scenario. (Consult [2] for the basics of how to perform the above two steps.) Of course, as I well knew, the only system RF losses for which that the basic Friis equation itself accounted was free-space loss (FSL). As is always the case, there were going to be other signal losses. Because both the radiated RF transmit and receiving station were placed on the desert earth ground, the question became, "What other significant RF losses, besides free space loss, would be experienced during the radiated RF outdoor test?"

I discovered through the reading of various technical RF literature that the losses that would have the greatest bearing on the received VHF-band signal test results would be:

- Multipath loss (MPL) [2];
- Fresnel zone loss (FZL) [2].

Illustrations of each are given in Figures 16.9 and 16.10.

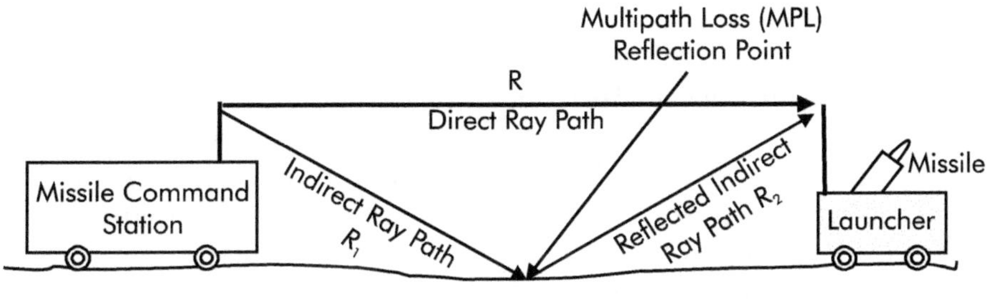

Figure 16.9 Outdoor RF test scenario illustrated accounting for MPL.

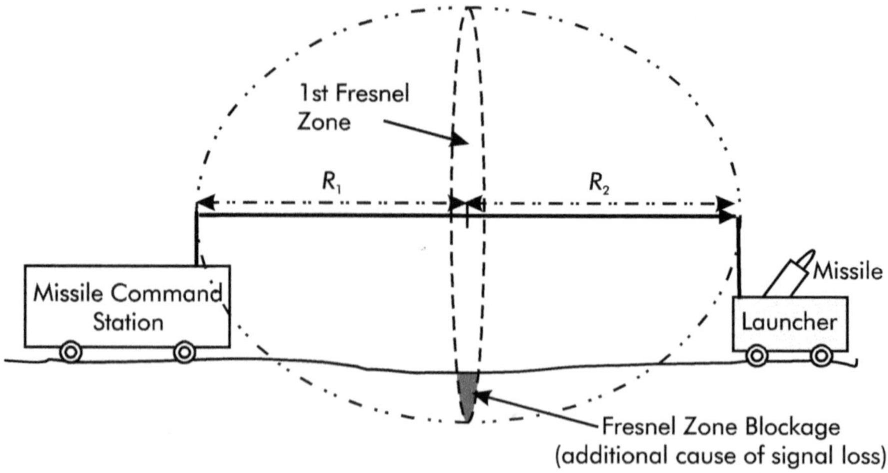

Figure 16.10 Outdoor RF test scenario illustrated accounting for FZL.

Were there other system losses that could be accounted for? Certainly. Transmission line losses are one. However, those losses would not be high enough to make a significant difference in the test result.

In the end, my numerical evaluation, when compared to the test results, came within 0.5 dB of the test results. That was so cool. If I had a football in my hand, I would have spiked it and danced in a little circle in my office.

16.3.1.3 The Happy Result

When my conclusion was reported to the US Army project office, the civilian and military staff were so pleased, they came over to the company office both to meet, congratulate, and thank me. What's more is that this success caused me to rise in technical prominence both within the company and at the US Army customer project office. From that day forward I was invited to other meetings in which RF radiation testing possibilities were discussed.

The importance of this above-described outcome is to understand that you cannot only rise in technical prominence in an engineering firm but you can also escape periods of unexpected company downsizing. When executives see your value to the company by way of the reputation you have built, you stay employed.

Unfortunately, at the end of November 1992, the company found it necessary to conduct a workforce reduction (called a RIF, i.e., an official term for layoff). However, to the gratitude of both me, my wife, and our three children, I was retained by CAS, Inc. until I left in February of 1996 to go to Texas Instruments for even greener pastures for my family and I.

16.3.2 B52H ARC-210 Warrior Radio VHF Max Range (300 nmi) In-flight Test

16.3.2.1 Introduction

In approximately January 2011, I received a phone call at my desk from a Boeing-Wichita project manager. This manager was having an unresolvable engineering problem. Responsible engineers were not able to show evidence from flight testing that a certain newly-installed VHF-band radio system could both establish and maintain in-flight VHF communications at the customer-required range of 300 nmi.

Three flight tests had already been conducted. Each yielded no positive outcome. I then was invited to a meeting in which the situation was carefully outlined. At that meeting I was requested to come and help solve their VHF engineering problem.

There is an underlying background story to this adventure.

16.3.2.2 Terms/Conditions of Service That Had To Be Outlined

A number of the engineering colleagues in the meeting (not the program manager, mind you; she was a good friend) were resentful toward me because of my past success(es) solving EM problems for their team that they had not been able to solve themselves. Simply put. . . a case of professional jealousy.

Nonetheless, the requesting manager and I were good friends. She knew my technical capability as an RF engineer quite well (that is, from the RF test set case previously described). She was the one that summoned me to the meeting to help. After hearing the totality of the engineering problem, I then proceeded to lay down, in no uncertain words, the terms and conditions for my joining this effort.

I told them that I had to have complete freedom and total control/authority over every technical aspect of this undertaking (I felt I had to do this because of a few of the curmudgeon personalities at the meeting). I further said that I had no problem with people asking questions.

Anyone could freely ask any question they wanted. But, I said there was to be no foolish, superfluous arguing with me over any step that I chose to undertake to solve the problem. I said that, if what I asked during this effort was not followed, I would leave their project to them and go back to the current RF-related program that I was on (on which I was having considerable technical success). Aloud for all to hear, the program manager granted that freedom to me, and my terms and conditions of service were met. We then proceeded forward.

What did they have to lose now? More time and more unplanned expense. Of course, I was going to be granted my freedom. They were desperate. Lots of pressure was applied from company leadership to complete this 300-nmi requirement verification effort and close the program out. Delays were costing considerable money. That was that.

16.3.2.3 Three Previous Flight Test Failures

After reviewing the past efforts undertaken, what became immediately apparent was that the flight test failures were the simple case of a lack of RF technical knowledge on terrestrial ground-based VHF-band wave propagation. Peculiar to the failures were specific RF losses not being accounted for in that flight test, making in-flight two-way VHF communication impossible. Past team members simply did not understand the negative contribution that Fresnel zone blockage was having on their flight test.

By flying the bomber over the Nevada mountain range, the on-foot person carrying the manpack VHF transceiver was on a mountain side opposite to the airplane flying overhead. See Figure 16.11. As a result, the mountain in which the man was standing completely blocked the receiving VHF signal and the signal was not being received by the VHF transceiver manpack.

Thus, the flight test plan and procedure needed to be revised. I will address that soon.

If technical terms such as MPL or Fresnel zone blockage are not familiar to the reader, it would be well to consult my first published technical Artech House book as documented in [2]. In that book, go to Chapter 4 for technical discussion on MPL and Chapter 5 for FZL.

The mountains that were completely blocking off the radiated VHF signal in all three failed flight test events needed to no longer

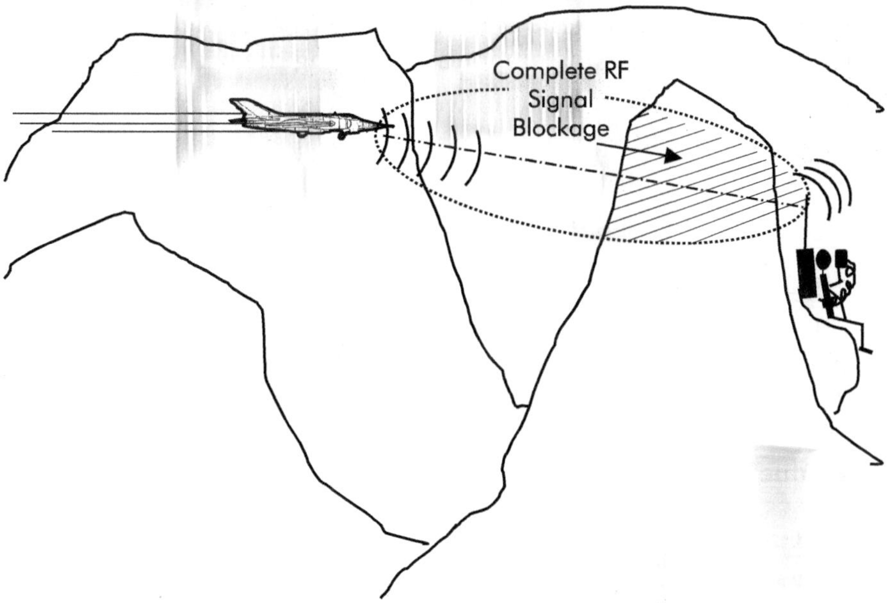

Figure 16.11 Outdoor RF test scenario illustrated showing the effect of FZL (i.e., Fresnel zone blockage).

have an impact, and I proposed an altogether different scenario for flight test event 4. I was thoroughly convinced that this would be the flight test that would lead to complete technical success, verify the requirement, and close out the program for the company.

16.3.2.4 New Flight Test Scenario

I proposed to the program manager that the site of the next flight test event be moved to a location close to the Pacific Ocean; that is, Vandenberg Air Force Base (AFB) in California, completely out of the Nevada mountain ranges. I further directed for the new test scenario that:

1. The person donning the VHF manpack radio would station himself on the side of a west-facing mountain, that is, one that faced the Pacific Ocean (no obstructions);
2. The test aircraft during flight test would:
    - Fly east-to-west, that is, toward the Pacific Ocean;

- Fly directly overhead of the man carrying the VHF manpack radio on the side of mountain facing the ocean;
- Continue to fly east to west out to sea to a ground distance of 340 nmi;
- Fly at an altitude of 35,000 ft when directly overhead of the man donning the manpack VHF radio.

See Figures 16.12 and 16.13.

By using this new test scenario, there would be absolutely no blockage of the Fresnel zone that would form between the test plane VHF antenna and the mountain-stationed VHF manpack radio. Furthermore, the only major RF loss would be the MPL that would take place due to the ocean-reflective indirect antenna ray coming from

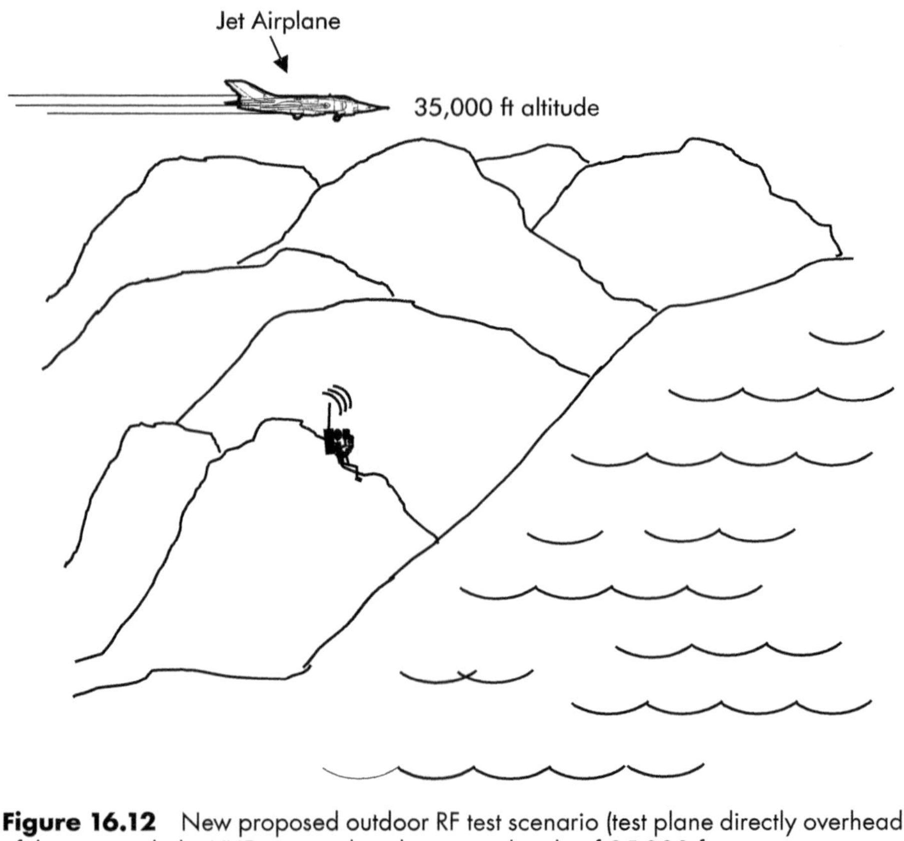

Figure 16.12 New proposed outdoor RF test scenario (test plane directly overhead of the man with the VHF manpack radio at an altitude of 35,000 ft.

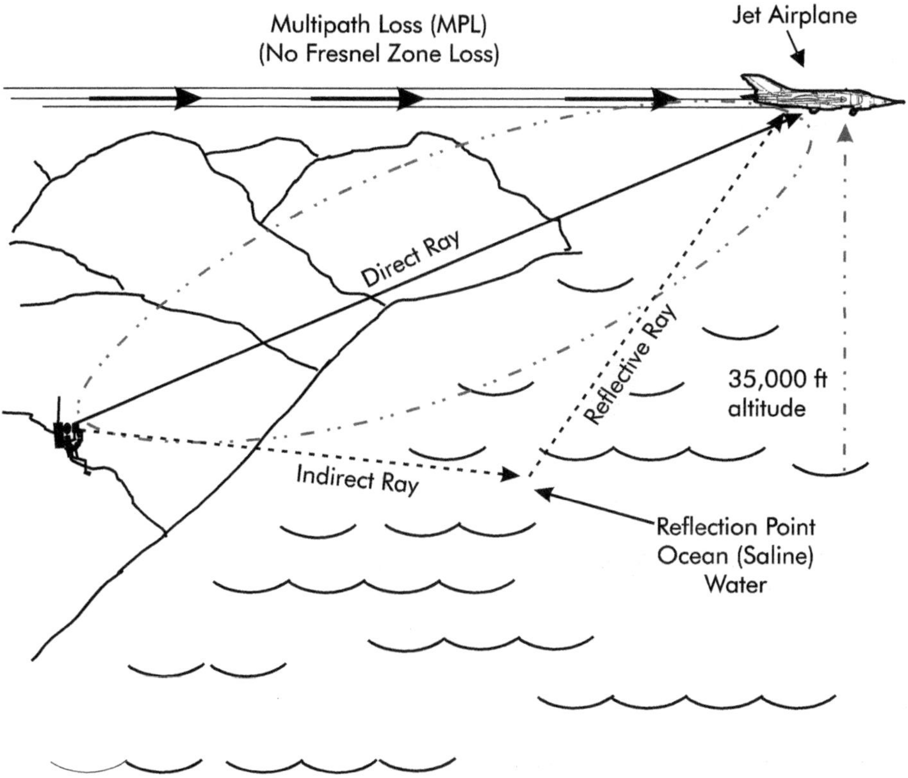

Figure 16.13 New proposed outdoor RF test scenario (test plane at 35,000 ft and 340 nmi out to sea).

the test plane and would go to the VHF manpack radio (and vice versa in the return transmitting direction). This new flight test geometry would minimize MPL as much as could be done [2]. The results of the flight test that used the new flight test scenario are described below.

16.3.2.5 Test Success: Successful Verification of 300-nmi Range Capability of the Newly Installed VHF Radio

What was reported by the stationary mountain-located VHF manpack radio man and the east-to-west flying bomber test airplane crew was that successful radiated VHF radio communication occurred even beyond 300 nmi, both going away from shore and coming back to it. Successful communications began from the time the test aircraft was directly overhead of the mountain-stationed manpack radio (i.e., 0 nmi ground distance) to when the test airplane was 300 nmi ground distance and beyond.

In-flight VHF communication checks were made every 20 nmi starting at a ground distance of 240 nmi away from the mountain until a 340 nmi ground distance was reached. Both the crew man on the mountain and the test airplane VHF communication airmen reported that VHF-band communications were "loud and clear" during the entire test flight and across the full VHF communications distance.

16.3.2.6 Why Test Scenario Achieved Immediate Success

The reason for the successful final flight testing was two-fold.

1. Flight test occurrences of Fresnel zone blockage were no longer occurring. No mountains were in the path of the line-of-sight (LOS). The ocean shoreline geography offered no obstruction whatsoever.
2. Whereas there was no occurrence of MPL for the first three flight test events (i.e., due to the fact that the test plane was flying in the Nevada desert mountains) there was only a small and numerically predictable occurrence of MPL for the case of the overhead test plane flying east-to-west at 35,000 ft out to sea.

To the relief and great joy of the program manager, the test result was successful with no repeats of flight testing being needed. The customer-requested VHF radio range of 300 nmi communication range could be supported. As a result, the successful result enabled the engineering program to be completely closed out for that Boeing program manager.

16.4 LESSONS LEARNED

What is important to learn from each of the described instances of technical problems is that, first, a thorough understanding of RF physics and propagation should be secured in one's own mind before any mitigation steps are attempted. When you have sufficient knowledge and understanding of the situation and a sound perception of the RF principles involved in a scenario, then you have a chance of resolving the problematic dilemma at hand. I hope from this second edition that this immutable truth has been sufficiently demonstrated.

I want to encourage the reader, young or old, to go forth and boldly resolve the arising RF-related problems you face now and will face at work. If you do that, you will become deservedly famous in your company and be among those that can be counted upon to help solve technical problems/situations in your engineering firm.

16.5 FINAL REMARKS TO THE READER

In my final remarks for this second edition, there is a book I highly recommend the reader to read [9]. It is not a textbook, and I still read it often. The book is called *The Road Less Traveled*, written by the late Dr. M. Scott Peck. The book is a priceless gem.

I read it frequently between the years of 1990 to 2020, that is, until I retired in June 5, 2020. It is not only very well written but contains four timeless principles that are, in my estimation, invaluable for the engineering workplace. I would encourage engineers to keep and follow them for the remainder of one's professional life and beyond [3].

In essence, Dr. Peck says that "Life is difficult. This is a great truth." My fellow engineering professional, electromagnetic engineering life work is also difficult. To be successful in it, one must approach engineering challenges by following Peck's four principles of personal discipline [3]:

1. Accept responsibility;
2. Delay gratification;
3. Be dedicated to truth;
4. Live a balanced life.

I will elaborate on each of these from my own personal experience and perspective. The reader, of course, is welcome is to draw his/her own conclusions whatever they may be.

Accept Responsibility

As I wrote this book, I made the first assumption that you, the reader, are out of college and out of engineering school. The second one I made was that you have accepted a position as an electromagnetic (or other discipline) engineer in an engineering firm. Under both these assumptions, let me state what I have found to be an immutable fact.

Your professors will not go with you into the workplace. You are essentially on your own to be the expert. Therefore, continue to learn, learn, learn. You have senior colleagues that will be with you. Regardless, you were hired by your new engineering firm because your recognized college GPA gave evidence that you had the "smarts" and the perseverance to finish obtaining your BS engineering degree. That already speaks highly of you. The same determination that got you to complete your course work and walk across the university graduation stage will be needed in the engineering workplace.

Know that you will encounter several truly challenging or sometimes even seemingly insurmountable engineering problems. My advice—don't run from the problems.

Embrace them.

Accept the situation gladly. Calmly resolve the problem situation. You are the paid engineering expert now. Accept the honor that has been bestowed upon you. Be true to the trust that you have been given.

Have faith in yourself. Assure yourself that you will find the answer. With this resolve you cannot help but succeed.

Delay Gratification

Some people have the tendency to start on easy problems first and then move over to the difficult ones later. That is, of course, one approach. I call this approach the "picking the low-hanging tree fruit." Problems, regardless of how great or small, require being solved. Solving those problems is how engineering companies make their money, enabling them to pay engineering employees like you. So, I say start on the difficult engineering problems first as soon as possible. The sooner you get started, the quicker you will find the solution. Also, always be aware that there will always be another engineering problem to solve on another day. Get used to it.

Be Dedicated to Truth

To me, being dedicated to truth in the workplace simply means never deceiving yourself into continuing to follow an approach that isn't working. If your first approach isn't working, stop doing it. Try a different approach. Be humble enough to stop and consider another alternative. Remember that you are trying to get to the source of the problem. This means being dedicated to pursuing the right path no matter where it leads.

Lead/Live a Balanced Life

This last principle is one I discovered later in life, but it has become as important to me as the previous three. Living a balanced life to me means, while giving your all to solving the engineering problem during the workday, don't neglect to do those things you really enjoy doing off the job. For me these things include hiking, backpacking, camping, woodworking, rifle and shotgun shooting, gun collecting, and going to have lunch with your friends, having evening dinner out with your wife, friends, or grown children, and spending time with your children/grandchildren. These are the things that are super important to me. What are those things that are super important to you?

Find those things that help "sharpen your intellectual knife." You will need to sharpen your knife again and again and again.

My best wishes to you in all your future endeavors.

References

[1] Friesen, D. J., *Electromagnetics For Engineers, Volume 1: Electrostatics and Magnetostatics*, Norwood, MA: Artech House, 2023.

[2] Friesen, D. J., *Introduction to Antennas and RF Propagation Analysis*, Norwood, MA: Artech House, 2022

[3] Peck, M. Scott, *The Road Less Traveled*, New York: Touchstone, 1978.

A

SCIENTIFIC PREFIXES

Prefix	**Abbreviation**	**Decimal Equivalent**
Exa	E	10^{18}
Peta	P	10^{15}
Tera	T	10^{12}
Giga	G	10^{9}
Mega	M	10^{6}
Kilo	k	10^{3}
Centi	c	10^{-2}
Milli	M	10^{-3}
Micro	m	10^{-6}
Nano	n	10^{-9}
Pico	p	10^{-12}
Femto	f	10^{-15}
Atto	a	10^{-18}

B

SCIENTIFIC CONSTANTS

e = charge of an electron = 1.602×10^{-19} C

h = Planck's constant = 6.626×10^{-34} Js

k = Boltzmann's constant = 1.380×10^{-23} J/K

T_0 = standard (noise) temperature = 17°C = 290 Kelvin (or K)

$c = v_c$ = velocity of light in a vacuum = 299,792,500 +300 m/s

e = base of natural logarithm system = 2.718

η = free space characteristic impedance – $120\pi\ \Omega = 377\Omega$

ε_0 = dielectric permittivity of free space = 8.854×10^{-12} F/m

μ_0 = magnetic permeability of free space = 1.257×10^{-6} H/m

C

RULES BY WHICH TO PERFORM VECTOR ANALYSIS

C.1 PARALLELOGRAM RULE

Step 1. Parallelogram rule: Connect vectors **A** and **B** in Figure C.1(a) at the tail side of each vector in Figure C.1(b). Connect a third vector **C** tail at the location given in Figure C.1(b). Draw a line parallel to vector **A**, and a second line parallel to vector **B** as shown in Figure C.1(b).

C.2 HEAD-TO-TAIL RULE

Step 1. Head-to-tail rule (**A** + **B**): Examine the lower half of Figure C.1(b) and note vectors **A**, **B**, and **C** connected as depicted in Figure C.2(b).

Step 2. Head-to-tail rule (**B** + **A**): Examine the upper half of Figure C.1(b) and note vectors **A**, **B**, and **C** connected as depicted in Figure C.2(b).

Figure C.1 (a): Two vectors, **A** and **B** (b): parallelogram rule.

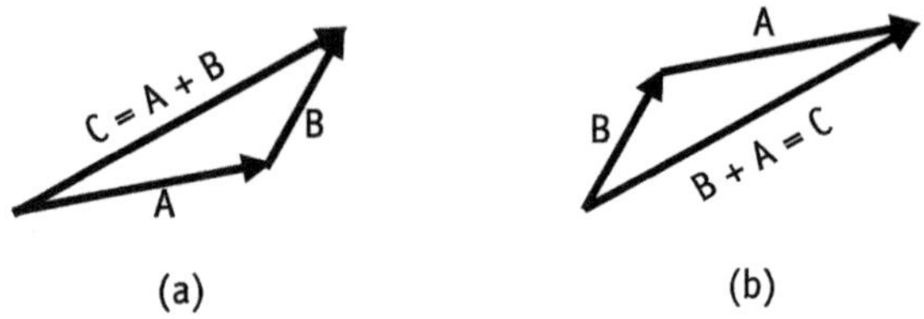

Figure C.2 Head-to-tail rule: (a) Head-to-tail rule, **A** + **B**, (b) head-to-tail rule, **B** + **A**.

Step 3. Perform the respective vector addition as depicted in Figures C.2(a, b).

D

ELECTROMAGNETIC SPECTRUM AND FREQUENCY BAND DESIGNATIONS

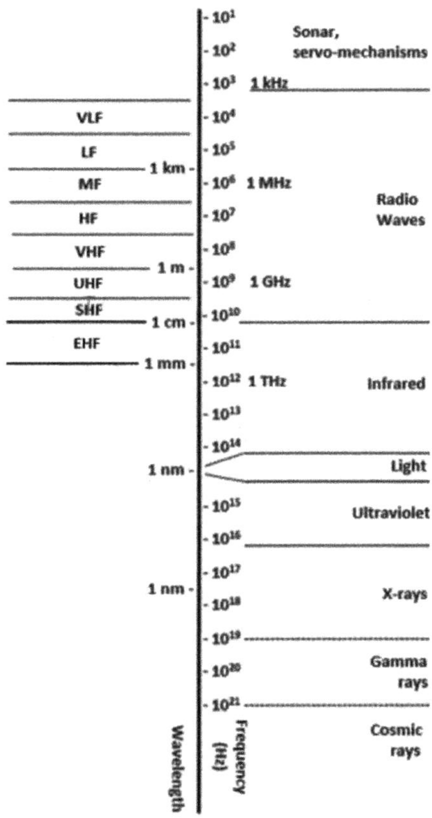

Figure D.1 Electromagnetic spectrum.

Table D.1
Frequency Band Designations

Band Name	Frequency Range
UHF	0.3–1.0 GHz
L-band	1.0–1.5 GHz
S-band	1.5–3.9 GHz
C-band	3.9–8.0 GHz
X-band	8.0–12.0 GHz
Ku-band	12.5–18.0 GHz
K-band	18.0–26.5 GHz
Ka-band	26.5–40.0 GHz
U-band	40.0–60.0 GHz
V-band	50.0–75.0 GHz
W-band	75.0–100.0 GHz

E

TRANSMISSION LINE EQUATIONS, GENERAL LINE EXPRESSIONS, AND IDEAL LINE EXPRESSIONS

Table E.1
Transmission Line Equations, General Transmission Line Expressions, and Ideal Line Expressions

Quantity	General Line Expression	Ideal Line Expression
Propagation constant	$\gamma = \alpha + j\beta$	$\gamma = j\omega(LC)^{1/2}$
Phase constant, β	Imaginary Part of γ	$\beta = \omega(LC)^{1/2} = 2\pi/\lambda$
Attenuation constant, α	Real Part of γ	0
Characteristic impedance, Z_0	$Z_0 = \left[\frac{(R + j\omega L)}{(G + j\omega C)}\right]^{1/2}$	$Z_0 = \left(\frac{L}{C}\right)^{1/2}$
Input impedance	$Z_{in} = Z_0\left[\frac{(Z_L + Z_0 \tanh(\gamma l))}{(Z_0 + Z_L \tanh(\gamma l))}\right]^{1/2}$	$Z_{in} = Z_0\left[\frac{(Z_L + jZ_0 \tan(\beta l))}{(Z_0 + jZ_L \tan(\beta l))}\right]^{1/2}$
Impedance of short-circuited line, $Z_L = 0$	$Z_{s.c.} = Z_0 \tanh(\gamma l)$	$Z_{s.c.} = jZ_0 \tan(\beta l)$
Impedance of open-circuited line, $Z_L = \infty$	$Z_{o.c.} = Z_0 \coth(\gamma l)$	$Z_{o.c.} = -jZ_0 \cot(\beta l)$
Impedance of line, odd no. of ½ λs long	$Z = Z_0\left[\frac{(Z_L + Z_0 \coth(\alpha l))}{(Z_0 + Z_L \coth(\alpha l))}\right]$	$Z = \frac{Z_0^2}{Z_L}$
Impedance of line, integer no. of ½ λs long	$Z = Z_0\left[\frac{(Z_L + Z_0 \tanh(\alpha l))}{(Z_0 + Z_L \tanh(\alpha l))}\right]$	$Z = Z_L$
Voltage reflection coefficient	$\Gamma = \frac{(Z_L - Z_0)}{(Z_L + Z_0)}$	$\Gamma = \frac{(Z_L - Z_0)}{(Z_L + Z_0)}$

Table E.2
Helpful Transmission Line-Related Equations

Equation	Explanation
$r = \frac{1+\lvert\Gamma\rvert}{1+\lvert\Gamma\rvert}$ $\Gamma = \frac{(r-1)}{(r+1)}$	R = VSWR $\lvert\Gamma\rvert$ = magnitude of reflection coefficient
$\Gamma = \frac{(R-Z_0)}{(R+Z_0)}$ $r = \frac{R}{Z_0}$ $r = \frac{Z_0}{R}$	Γ = reflection coefficient (real) at a point in a line where impedance is real (R) $R > Z_0$ (at voltage maximum) $Z_0 < R$ (at voltage minimum)
$\frac{P_t}{P_i} = \lvert\Gamma\rvert^2 = \left(\frac{r-1}{r+1}\right)^2$ $\frac{P_t}{P_i} = 1-\lvert\Gamma\rvert^2 = \frac{4r}{(r+1)}$	P_t = transmitted power P_r = reflected power P_i = incident power

F

MAXWELL'S EQUATIONS

F.1 FREE SPACE SET

Table F.1

Maxwell's Equations, Free Space Set: Point Form Versus Integral Form

Point Form	**Integral Form**
$\nabla \times \bar{\mathbf{H}} = \frac{\partial \bar{\mathbf{A}}}{dt}$	$\oint \bar{\mathbf{H}} \ast d\mathbf{l} = \int_S \left(\frac{\partial \bar{\mathbf{D}}}{\partial t}\right) \ast d\bar{\mathbf{S}}$
$\nabla \times \bar{\mathbf{E}} = -\frac{\partial \bar{\mathbf{B}}}{dt}$	$\oint \bar{\mathbf{H}} \ast d\mathbf{l} = \int_S \left(\frac{\partial \bar{\mathbf{D}}}{\partial t}\right) \ast d\bar{\mathbf{S}}$
$\nabla \ast \bar{\mathbf{D}} = 0$	$\oint_S \bar{\mathbf{D}} \ast d\bar{\mathbf{S}} = 0$
$\nabla \times \bar{\mathbf{B}} = 0$	$\oint_S \bar{\mathbf{B}} \ast d\bar{\mathbf{S}} = 0$

F.2 GENERAL SET

Table F.2

Maxwell's Equations, General Set: Point Form Versus Integral Form

Point Form	**Integral Form**	
$\nabla \times \bar{\mathbf{H}} = \bar{\mathbf{J}}_c + \dfrac{\partial \bar{\mathbf{D}}}{dt}$	$\oint \bar{\mathbf{H}} * d\mathbf{l} = \int_S \left(\dfrac{\partial \bar{\mathbf{D}}}{\partial t} \right) * d\bar{\mathbf{S}}$	(Ampere's law)
$\nabla \times \bar{\mathbf{E}} = -\dfrac{\partial \bar{\mathbf{B}}}{dt}$	$\oint \bar{\mathbf{H}} * d\mathbf{l} = \int_S \left(\dfrac{\partial \bar{\mathbf{D}}}{\partial t} \right) * d\bar{\mathbf{S}}$	(Faraday's law; *S* fixed)
$\nabla * \bar{\mathbf{D}} = \rho$	$\oint_S \bar{\mathbf{D}} * d\bar{\mathbf{S}} = \int_v \rho \, dv$	(Gauss' law)
$\nabla * \bar{\mathbf{B}} = 0$	$\oint_S \bar{\mathbf{B}} * d\bar{\mathbf{S}} = 0$	(Nonexistence of magnetic monopole)

ACRONYMS AND ABBREVIATIONS

A	Ampere
ABT	Air-breathing threat
AC	Alternating current
AFB	Air force base
Al	Aluminum
AL	Alabama
A/m	Amperes per meter
AWG	American Wire Gauge
BDS	Boeing Defense System
C	Coulomb
C/m	Coulomb per meter
CFRP	Carbon fiber reinforced plastic
CAS, Inc.	Clark and Stender Incorporated
CFR	Code of Federal Regulations
dB	Decibel
dc	Direct current
E-field	Electric field
ESD	Electrostatic discharge
EMC	Electromagnetic compatibility

EMF	Electromotive force
FL	Florida
F/m	Farads per meter
FSL	Free-space loss
IC	Integrated circuit
H-field	Magnetic field
Hz	Hertz
LNA	Low noise amplifier
M	Meter
m/s	Meters per second
MD	Medical doctor
nmi	Nautical mile
NGC	Northrop-Grumman Corporation
NM	New Mexico
PCB	Printed circuit boards
PD	Product development
PhD	Doctor of philosophy
PPO	Patriot Project Office
R&D	Research and development
SIL	System Integration Laboratory
s	Second
S	Siemens
S/m	Siemens per meter
TBM	Tactical ballistic missile
TTL	Transistor to transistor logic
US	United States of America (USA)
V/m	Volts per meter
WA	Washington
Wb	Weber
WSMR	White Sands Missile Range

LIST OF SYMBOLS

λ	Wavelength (m)
c	Speed of light, approximated to 300 × 10E6 m/s
f	Frequency (Hz or cycles/s)
T	Period (s)

SELECTED BIBLIOGRAPHY

Blake, L. V., *Antennas*, Norwood, MA: Artech House, 1984.

Brown, R. G., et al., *Lines, Waves, and Antennas: The Transmission of Electric Energy*, Second Edition, New York: John Wiley and Sons, 1973, Chapters 12–15.

Budlong, A. L., *The A. R. R. L. Antenna Book*, 9th Edition, West Hartford, CT: The American Radio Relay League, 1960.

Cheng, D. K., *Fundamentals of Engineering Electromagnetics*, Reading, MA: Addison-Wesley, 1993.

De Martino, A., *Introduction to Modern EW Systems*, Norwood, MA: Artech House, 2012, Appendix C.

Elbert, B. R., *The Satellite Communication Applications Handbook*, Norwood, MA: Artech House, 1997.

Friesen, D. J., *Introduction to Antennas and RF Propagation Analysis*, Norwood, MA: Artech House, 2022

Friesen, D. J., *Electromagnetics for Engineers: Electrostatics and Magnetostatics, Volume 1*, Norwood, MA: Artech House, 2022.

Hecht, E., *Schaum's Outline of Theory and Problems of Optics*, New York: McGraw-Hill, 1975.

Howard W. Sams and Co., *Reference Data for Radio Engineers*, 6th Edition, Indianapolis, IN: ITT, 1976.

Howard W. Sams and Co., *Reference Data for Radio Engineers*, 5th Edition, Indianapolis, IN: ITT, 1972.

Johnson, R. C. (ed.), *Antenna Engineering Handbook*, 3rd Edition, New York: McGraw-Hill, 1993.

Jordan, E. C., *Electromagnetic Waves And Radiating Systems*, London, Constable and Company, Ltd., 1953.

Kennedy, G., *Electronic Communications Systems*, 2nd Edition, New York: McGraw-Hill, 1977.

Macnamara, T., *Handbook of Antennas for EMC*, Norwood, MA: Artech House, 1995.

Peck, M. S., *The Road Less Traveled*, New York: Touchstone, 1978.

Poisel, R. A., *Antenna Systems and Electronic Warfare Applications*, Norwood, MA: Artech House, 2012.

Russer, P., *Electromagnetics, Microwave Circuit and Antenna Design for Communications Engineering*, Norwood, MA: Artech House, 2006.

Saakian, A., *Radio Wave Propagation Fundamentals*, Norwood, MA: Artech House, 2011.

Shadowitz, A., *The Electromagnetic Field*, New York: Dover Publications, Inc., 1975.

Stutzman, W. L., and G. A. Thiele, *Antenna Theory and Design*, New York: John Wiley and Sons, 1981.

Volakis, J. L. (ed.), *Antenna Engineering Handbook*, 4th Edition, McGraw-Hill, 2007.

Warnick, K. F., and P. Russer, *Problem Solving in Electromagnetics, Microwave Circuit and Antenna Design for Communications Engineering*, Norwood, MA: Artech House, 2006.

Wiley, R. G., *Electronic Intelligence: The Analysis of Radar Signals*, 2nd Edition, Norwood, MA: Artech House, 1993, Chapters 4 and 5.

ABOUT THE AUTHOR

Dean Friesen graduated from Kansas State University (Manhattan, KS), College of Engineering in 1987 with a BS degree in engineering technology (electrical/electronic emphasis) and in 1984 with an associates of technology (AT) degree in electronic technology from Kansas Technical Institute (now Kansas State Polytechnic, Salina, KS). Mr. Friesen worked a total of 33 years in the EM-related engineering-related disciplines of EW, RF/MW/mm-wave, HEMP, EMC/EMI, E3, and EME before retiring from The Boeing Company on June 5, 2020 as the Boeing Commercial Airplanes, Product Development Org, EME and Antennas Group, senior principal investigator (June 2016 to June 2020).

Mr. Friesen began his military defense engineering career at Boeing-Wichita (KS, B52 Defensive Systems) in June 1987 where he performed B52H Electronic Warfare work for the USAF for 2 and 1/2 years. In January 1990, he joined the International Space Station (ISS) project (formerly called Space Station Freedom) at Boeing-Huntsville (AL) to work for 1 and 3/4 years. Mr. Friesen's desire to return to US defense engineering work moved him to join CAS, Inc (Patriot Weapon System Analysis, Huntsville, AL) in October 1991. He also worked for TX Instruments (defense sector, RF/MW Group, Dallas, TX) in February 1996, Northrop-Grumman Corporation (Land Combat Systems (LCS), Huntsville, AL) in June 1999, and Boeing-Wichita (Boeing Defense Systems) again in October 2003.

The untimely closing of the Boeing-Wichita site in 2012 caused Mr. Friesen to join BCA to work in the Puget Sound area (Everett/Renton/Mukilteo) in Washington state for a total of 8 years. He worked the first four of those years as the 737MAX airplane BCA EME team engineering lead. In September 2016, he was recruited specifically to join the BCA Product Development organization, EME and Antennas Group, as a principal investigator where he was promoted in 3 months to be the group's sole senior principal investigator (4 years) before retiring June 2020. Mr. Friesen and his wife now reside in Huntsville, AL.

INDEX

A

B

D

E

F

Z

Artech House
Electromagnetic Analysis Library

Christos Christodoulou, Series Editor
Vince Rodriguez, Series Editor

Advanced Computational Electromagnetic Methods and Applications, Wenhua Yu, Wenxing Li, Atef Elsherbeni, Yahya Rahmat-Samii, editors

Advanced FDTD Methods: Parallelization, Acceleration, and Engineering Applications, Wenhua Yu, et al.

Advances in FDTD Computational Electrodynamics: Photonics and Nanotechnology, Allen Taflove, editor

Analysis Methods for Electromagnetic Wave Problems, Volume 1, Eikichi Yamashita, editor

Analysis Methods for Electromagnetic Wave Problems, Volume 2, Eikichi Yamashita, editor

Analytical and Computational Methods in Electromagnetics, Ramesh Garg

Analytical Modeling in Applied Electromagnetics, Sergei Tretyakov

Anechoic Range Design for Electromagnetic Measurements, Vince Rodriguez

Applications of Neural Networks in Electromagnetics, Christos Christodoulou and Michael Georgiopoulos

CFDTD: Conformal Finite-Difference Time-Domain Maxwell's Equations Solver, Software and User's Guide, Wenhua Yu and Raj Mittra

Chipless RFID Printing Technologies, Santanu Kumar Behera and Durga Prasad Mishra

Computational Electrodynamics: The Finite-Difference Time-Domain Method, Second Edition, Allen Taflove and Susan C. Hagness

Computational Electromagnetics, Prabha Umashankar and Allen Taflove

Electromagnetic Fields in Multilayered Structures: Theory and Applications, Arun Bhattacharyya

Electromagnetic Modeling of Composite Metallic and Dielectric Structures, Branko M. Kolundzija and Antonije R. Djordjevic

Electromagnetics for Practicing Engineers, Second Edition, Dean James Friesen

Electromagnetic Scattering and Material Characterization, Abbas Omar

Electromagnetic Waves in Chiral and Bi-Isotropic Media, I. V. Lindell, et al.

Electromagnetic Diffraction Modeling and Simulation with MATLAB®, Gökhan Apaydin and Levent Sevgi

Engineering Applications of the Modulated Scatterer Technique, Jean-Charles Bolomey and Fred E. Gardiol

Far-Field Wireless Power Transfer and Energy Harvesting, Naoki Shinohara and Jiafeng Zhou, editors

Fast and Efficient Algorithms in Computational Electromagnetics, Weng Cho Chew, et al., editors

FDTD Modelling of Metamaterials: Theory and Applications, Yang Hao and Raj Mittra

Generalized Multipole Technique, Christian Hafner

Grid Computing for Electromagnetics, Luciano Tarricone and Alessandra Esposito

High Frequency Electromagnetic Dosimetry, David A. Sánchez-Hernández, editor

High-Power Electromagnetic Effects on Electronic Systems, D. V. Giri, Richard Hoad, and Frank Sabath

Integral Equation Methods for Electromagnetics, Joseph R. Mautz and Nagayoshi Morita

Intersystem EMC Analysis, Interference, and Solutions, Uri Vered

Introduction to Electromagnetic Wave Propagation, Paul Rohan

Introduction to the Uniform Geometrical Theory of Diffraction, Derek A. McNamara, C. W. I. Pistorius, and Jan Malherbe

Iterative and Self-Adaptive Finite-Elements in Electromagnetic Modeling, Magdalena Salazar-Palma, et al.

Machine Learning Applications in Electromagnetics and Antenna Array Processing, Manel Martínez-Ramón, Arjun Gupta, José Luis Rojo-Álvarez, and Christos Christodoulou

New Foundations for Applied Electromagnetics: The Spatial Structure of Fields, Said Mikki and Yahia Antar

Numerical Analysis for Electromagnetic Integral Equations, Karl F. Warnick

Parallel Finite-Difference Time-Domain Method, Wenhua Yu, et al.

Practical Applications of Asymptotic Techniques in Electromagnetics, Francisco Saez de Adana, et al.

A Practical Guide to EMC Engineering, Levent Sevgi

Quick Finite Elements for Electromagnetic Waves, Second Edition, Giuseppe Pelosi, Roberto Coccioli, and Stefano Selleri

The Scattering of Electromagnetic Waves from Rough Surfaces, Peter Beckman and Andre Spizzichino

Shipboard Electromagnetics, Preston Law

Substrate Integrated Waveguides, Soumava Mukherjee, Prasun Chongder, and Animesh Biswas

Understanding Electromagnetic Scattering Using the Moment Method: A Practical Approach, Randy Bancroft

Wavelet Applications in Engineering Electromagnetics, Tapan K. Sarkar, Magdalena Salazar-Palma, and Michael C. Wicks

For further information on these and other Artech House titles, including previously considered out-of-print books now available through our In-Print-Forever® (IPF®) program, contact:

Artech House
685 Canton Street
Norwood, MA 02062
Phone: 781-769-9750
Fax: 781-769-6334
e-mail: artech@artechhouse.com

Artech House
16 Sussex Street
London SW1V 4RW UK
Phone: +44 (0)20 7596 8750
Fax: +44 (0)20 7630 0166
e-mail: artech-uk@artechhouse.com

Find us on the World Wide Web at: www.artechhouse.com